Practical Laser Safety

SECOND EDITION, REVISED AND EXPANDED

OCCUPATIONAL SAFETY AND HEALTH

A Series of Reference Books and Textbooks
on Occupational Hazards ● Safety● Health●
Fire Protection ● Security ● and Industrial Hygiene

Series Editor
ALAN L. KLING
Loss Prevention Consultant
Jamesburg, New Jersey

Additional Volumes in Preparation

Practical Laser Safety

SECOND EDITION, REVISED AND EXPANDED

D. C. Winburn

CONSULTANT
LOS ALAMOS, NEW MEXICO

CRC Press

Taylor & Francis Group

Boca Raton London New York

CRC Press is an imprint of the
Taylor & Francis Group, an **informa** business

CRC Press
Taylor & Francis Group
6000 Broken Sound Parkway NW, Suite 300
Boca Raton, FL 33487-2742

First issued in paperback 2019

ISBN-13: 978-0-8247-8240-5 (hbk)
ISBN-13: 978-0-367-40328-7 (pbk)

Library of Congress Cataloging-in-Publication Data

Winburn, D. C.
    Practical laser safety / D.C. Winburn.
        p.   cm. —— (Occupational safety and health ; 11)
    ISBN 0-8247-8240-2
    1. Lasers–Safety measures. I. Title. II. Series: Occupational
safety and health (Marcel Dekker, Inc.) ; 11.
    TA1677.W56   1989
    621.36'6'0289--dc20   89-27765
                                                             CIP

Visit the Taylor & Francis Web site at
http://www.taylorandfrancis.com

and the CRC Press Web site at
http://www.crcpress.com

To Dr. Keith Boyer, pioneer of laser fusion and leader of laser research at the Los Alamos National Laboratory for many years, for his support of my safety efforts,

and

To Yvonne, my bride of 45 years, for sharing my life for some 15 years with a mistress, the laser.

Preface

This second edition of *Practical Laser Safety* contains the latest revision of the American National Standards Institute's ANSI Z136.1 "Safe Use of Lasers." The Institute has formed two subcommittees: one for developing a standard for safety in health care facilities and one for developing a standard for fiber optics safety. These have established standards as ANSI Z136.2 and ANSI Z136.3, respectively. As in the first edition of *Practical Laser Safety*, which used the ANSI Z136.1 standard, portions of these three current standards that require compliance (of those entities adopting the standards) have been extracted with approval of the publisher, the Laser Institute of America. A critique of these standards follows the extractions.

An especially valuable addition to this second edition is the presentation of damage threshold values of lens materials used in laser-protective eyewear. Of special interest, too, is the discussion of the electrical hazards associated with lasers, which has been expanded.

This book was written for those individuals who are entering the field of laser technology, either as users of this magnificent tool, or as ones who have been assigned the task of controlling laser hazards by administrating a safety program. It is not intended to be a technical reference for all aspects of laser hazards, but rather a guide in applying fundamental principles of good laser safety practices.

My concern with the lack of practical information for lay people who are getting into laser applications was enhanced by the many inquiries I received during my six years as Secretary of the Laser Institute of America (1974-1979). Pilots, prospectors, welders, sign makers, laser-gun manufacturers, dentists, doctors, artists, musicians, and teachers with questions about this tool of an emerging technology helped decide this method of communicating some of my thoughts. My retirement after 10 years of deep involvement in providing a safe environment for laser users at the Los Alamos National Laboratory has offered me this opportunity.

Many users of lasers, or those responsible for safety programs involving a laser environment, do not desire an extensive understanding of the physics and mathematics associated with lasing conditions or laser characteristics, nor are they required to have an in-depth knowledge of biological damage mechanisms. However, adherence to regulations and standards is necessary, so an understanding of specific requirements is paramount, and considerable space has been allocated for this topic, especially with regard to details of the American National Standards Institute's document, ANSI Z136.1 "Safe Use of Lasers."

The eye is the organ most susceptible to damage from laser beams; therefore, protective eyewear is the dominant feature of a laser safety program. Selection of the appropriate type of eyewear for protection of laser users has been my main interest. My efforts in working with industrial firms over the past 15 years have resulted in the adapting of special glass filters to eye-protective devices that can be used for specific single-beam, multiple-beam, or broadband laser-beam protection. Eye protection is discussed in detail, with many specific applications.

As a representative of the Los Alamos National Laboratory, I served 10 years on the ANSI Z136.1 Committee. As a practicing Laser Safety Officer at the Los Alamos National Laboratory in the Laser Research and Technology Division for 10 years (1972-1982), without an incident of permanent biological damage, it has been my experience that, with proper indoctrination and continuing educational presentations, a successful program can be developed for laser users. An understanding of the potential hazards and their control must be an integral part of routine laser operations, and this aspect of a good safety program is stressed in this volume.

All safety engineers are taught that the causes of accidents are (1) unsafe acts and/or (2) unsafe conditions. An individual can be taught the principles of hazard control and personal attitude—and a trained Safety Officer (as required by ANSI Z136.1) can provide safe conditions in the laser environment. However, both the laser user and Laser Safety Officer must understand reasonable control procedures and be comfortable working with this valuable technology. That is my goal in writing this book: *practicality* in laser safety.

D. C. Winburn

Contents

Practical Laser Safety

SECOND EDITION, REVISED AND EXPANDED

1

Characteristics of Lasers

Lasers come in many sizes, shapes, and forms. Some lasers are as small as a grain of sand, while some large systems cover more than an acre of floor space when completely assembled. Certain lasers require toxic reactive materials, yet others need only a source of 110 V electricity (Fig. 1.1). Beams produced by some lasers have circular cross sections while other forms of beams include several small beams clustered together. Most lasers are mass produced, or "off-the-shelf," and are readily available from manufacturers, but many are "built from scratch" for special applications. The variety of lasers appears to be limited only to the imagination of the experimentor and the application in mind.

Beginning as a scientific curiosity in the early 1960s, laser energy absorption has found a multitude of applications, and the technology is considered just "emerging." This magnificient tool, in little more than 20 years, has revolutionized the industrial, medical, research, and art worlds. Lasers are now used by engineers, scientists, doctors, artists, and in seemingly countless other professions. Advantages of the laser have been realized by chemists, supermarket managers, printers, welders, construction contractors, computer manufacturers, surgeons, and others. Perhaps the biggest user of all is the military. Targeting, rangefinding, submarine tracking, and communications are among the many applications used in the military services. Probably the industry most affected by the laser is the communications field of fiberoptics, estimated to

Figure 1.1 Photograph of a simple laser system requiring only 110 V electrical source for activation. Students are shown in a classroom where a low-power HeNe laser is being used. (Courtesy of Hughes Aircraft, Carlsbad, California.)

be a three billion dollar industry by 1990. Not yet mentioned are uses in the fields of video discs, holography, entertainment, laser-fusion, metrology, spectroscopy, material cutting, and surveying. The ultimate use of laser beams is the death ray.

Laser rays can be controlled to extremely low powers to focus on microscopic spot sizes so that a 5-μm diameter hole can be "drilled" into one wall of a 150-μm diameter microballoon of glass, while other beams can be generated to powers of maximum possible output to be focused on fusion targets of the same size microballoons (Fig. 1.2). This versatility added to the other properties of coherent light make laser energy adaptable to a veritable plethora of applications. These characteristics include a variety of wavelengths (each having different qualities), the mode of propagation (either a continuous flow

Figure 1.2 Photograph of a portion of the world's largest CO_2 laser system, Antares, used in laser fusion research. (Courtesy of Los Alamos National Laboratory, Los Alamos, New Mexico.)

of photons or a burst of photons acting like a bullet of light, in a single pulse or in a series of pulses), focusability and dispersion (expansion) of the light rays through optical gear, absorption and transmission of the different wavelengths of various materials, and additional features newly discovered or being developed in physics and chemistry laboratories.

A few basic properties of the photons at hand need be known and understood to provide a safe laser environment and for the safe control of the potential hazards of laser light. These fundamental laser characteristics are described below.

WAVELENGTH

Because the wavelength of a laser beam determines its adsorption and transmission characteristics when interacting with materials, it is vital to know precisely the position of the wavelength of any laser beam in the electromagnetic

Figure 1.3 Representation of the electromagnetic spectrum that includes position of lasers.

spectrum. The wavelength produced by the lasing material of a laser is expressed in units of linear measurement as shown in the table in Figure 1.3. The wavelengths of essentially all lasers in common use are near the 1.0 micrometer (μm) spectral line, therefore, all units of laser wavelengths will be confined to the micrometer, or micron, as this unit is commonly known. That portion of the spectrum in which laser wavelengths are located is shown in Figure 1.4. Notice

Figure 1.4 Expanded portion of electromagnetic spectrum showing location of laser wavelengths.

that, in describing the various regions assigned to laser radiation, there are three categories:

1. Ultraviolet
2. Ocular focus
3. Infrared

Proliferation of laser technology has resulted in an increased number of non-technical users, therefore, it is logical to designate these spectral regions in simple terms. More sophisticated users will understand that in laser safety applications it is not necessary to use such terms as actinic ultraviolet (having photochemical properties) or near ultraviolet, or near infrared or far infrared. It is important, however, to know that the ocular focus region is a range of wavelengths from approximately 0.4 to 1.4 μm, which is focused by the eye's components with a power of approximately 100,000 times. (Components of the eye are described in Chap. 2.) Transmission of this portion of the spectrum through the eye results in these wavelengths being retinal hazards because these light rays reach the retina and are absorbed there. It is also important to know that this wavelength range, 0.4-1.4 μm, contains two distinct subranges. One range, 0.4-0.7 μm, is the *visible* portion of the spectrum and provides the light rays the eye uses to see. All wavelengths, or colors, varying from the blue colors near 0.4 μm, through the green and yellow colors near 0.5 μm, and including the orange and red colors from 0.6 to 0.7 μm are visible to the eye in this region. Transmission of light rays in the other range, 0.7-1.4 μm, also is permitted by ocular components except that the retina, which also absorbs this range of wavelengths, does not "see" them. In other words, this upper portion of the ocular focus region is *invisible*.

Owing to the focusability of some wavelengths by the eye as described above, these rays must be intercepted before they can cause harm to the retinal tissues, since even low energy laser beams, if concentrated by a factor of 100,000 (10^5), can cause damage to the eye as described in Chapter 3.

Wavelengths lower than 0.4 μm of interest in laser technology are located in the range of the spectrum designated as ultraviolet and do not pose as great a hazard to the eye because the outer components, especially the cornea, absorb essentially all of the rays' energies. The same is true of the portion of the spectrum called the infrared, which is above 1.4 μm. Ultraviolet and infrared wavelengths are not transmitted through the cornea and other exterior parts of the eye. These wavelengths are not focused on the retina, and, therefore, magnification of the rays' energies is not possible. Selection of protective eyewear should take this property of the nonfocusable laser beams into consideration as described in Chapter 8.

CONTINUOUS WAVE (CW) LASERS

Many applications of laser light require a steady stream of photons. These types of beams can be constructed of a variety of lasing materials; thus, the wavelengths available are extensive. Beam characteristics are readily measured because the beam is on continuously, reaching a constant condition, or steady-state. The laser beam diameter, for example, is a very important property in evaluating safety aspects of the beam and, with CW lasers, can be readily determined. The output power of a CW laser can be measured by simple instruments and is always presented in watts.

PULSED LASERS

Pulsed laser beams are actually "bullets" of light. Some pulses of laser light are created by chemical reaction, but usually the pulse is created by "chopping" a small portion of a CW beam either mechanically or by electronic methods. Terms used with pulsed mechanisms are choppers, Q-switched, and mode-locked. The descriptive units used to describe the length in time of a single pulse are listed as follows:

$$sec = second$$
$$msec = millisecond = .001 \, sec \qquad = 10^{-3} \, sec$$
$$\mu sec = microsecond = .000,001 \, sec \qquad = 10^{-6} \, sec$$
$$nsec = nanosecond = .000,000,001 \, sec \qquad = 10^{-9} \, sec$$
$$psec = picosecond = .000,000,000,001 \, sec = 10^{-12} \, sec$$

The definition of a pulsed laser per ANSI Z136.1 is one that delivers its energy in the form of a pulse or a train of pulses of less than 0.25 seconds per pulse. Most commercial pulsed lasers operate in the range of milliseconds to picoseconds. Only in the past several months have subpicosecond pulses been reported in research literature.

The energy of a pulsed laser is concentrated in a "bullet" of light. It damages biological tissue acoustically, that is, by a mechanical blast interaction. For example, a nanosecond pulse (10^{-9} sec) produces a 0.3 meter (approximately one foot) length of light traveling at the light speed of 3×10^8 meters/ sec. Consequently, low energies in the ocular focus region (0.4-1.4 μm) can produce retinal damage. All pulsed lasers emitting in the ocular focus wavelength band should be considered Class 4 in the ANSI Z136.1 classification scheme which is described in Chapter 6.

POWER AND POWER DENSITY OF BEAM

In evaluating a CW laser environment, it is essential to know the output power of the beam and the beam diameter in order to determine the watts per square centimeter (W/cm^2), the characteristic of the beam described as the beam "power density." The technical term for power density is *irradiance*, but power density is more descriptive and will be used throughout this book. Power densities may vary from a few W/cm^2 to hundreds of W/cm^2.

To obtain the power density of a beam, of course, the beam diameter must be known. For manufactured lasers it will be listed in the laser literature accompanying the equipment. Usually, the cross section of a beam is a simple circular area, the beam being a simple solid cylinder of light. However, if the diameter must be measured, technical assistance is required because most solid cross section cylindrical beams have a "gaussian" distribution which describes a beam of higher intensity near the center and lower intensity near the exterior. Some other cross sectional shapes are described in more technical publications listed in the references at the end of this chapter.

ENERGY AND ENERGY DENSITY OF BEAM

The watt (W) is the measurement of power in a CW laser beam. The joule (J) is the measurement of energy in a pulsed beam. The relationship between the watt and the joule is as follows:

1 joule (energy) = 1 watt (power) $\times$ 1 second (time)

or

$$1 \text{ watt} = \frac{1 \text{ joule}}{1 \text{ second}}$$

The energy intensity within a pulsed beam depends on the beam diameter in terms of square centimeters, just as described for the CW beam intensity. It is vital to know the beam diameter so that the joules per square centimeter, or energy density, can be determined. Technically, J/cm^2 is defined as *radiant exposure*, but in this book the pulsed beam intensity will be called the energy density.

BEAM DIVERGENCE

All laser beams have a tendency to increase in diameter as they propagate from the lasing chamber. This spreading out of the beam is called "beam divergence." Manufactured lasers list this property of the beam in the accompanying literature and state the divergence as a solid angle, Θ, expressed in milliradians. It is not necessary to understand the technical details of measuring divergence or the influence of this property in evaluating control of the beam's potential hazard except in applications that require long distances to perform specific functions. Lasers confined in a room or laboratory, unless they are plasma diodes or very special types, are not apt to have their beams increase appreciably in diameter from the beam output chamber to the target or beam stop. However, beam propagation over long distances in air can result in appreciable beam diameter increase. Military applications such as range-finding and target-sighting are examples of beam growth. Also, in transmitting beams through fiber-optics for long distances, "repeater" stations are required at intervals to renew the original beam characteristics.

It is important to realize in evaluating a laser environment that if the maximum available power density or energy density is used in allowing for eye protection, a safety factor automatically accompanies laser personnel at any point downstream from that maximum because the natural divergence of the beam results in the lowering of the beam intensity.

DISCUSSION OF BEAM REFLECTIONS

Laser beams are reflected to some extent from any surface contacted. If the reflecting surface is shiny like a mirror, the reflection is called "specular." If the reflecting surface is not shiny, the reflection is called "diffuse." The borderline between a specular reflection and a diffuse reflection is difficult to define. In fact, some shiny surfaces reflect less of the beam than some diffuse surfaces because some materials, such as clear glass, permit transmission of some wavelengths. In addition, some shiny surfaces, such as colored glasses or plastics, reflect less of the beam because some of it may be absorbed by the colored material. No surfaces except mirrors or other optics designed to be so, are precisely flat, therefore, essentially all reflections of laser beam result in spreading of the beam or beam divergence. This spreading of the beam is much more pronounced when reflected by a rough or diffuse-reflecting surface. For example, reflections from a specular surface are reduced only by a small percentage of the original beam from relatively flat surfaces, but reflections of laser beams from diffuse surfaces can be only a small fraction, hundredths or even thousandths of the original beam intensity. The color of the diffuse reflector is of less importance than (1) the roughness of the surface, (2) the distance the eye is from the reflecting surface, or (3) the reflection angle. A useful and widespread method of attenuating the beam before detection is to reflect it from a diffusely reflecting surface such as a refractory brick. Then, according to Lambert's law, the intensity of the reflected beam (E_r) is directly proportional to the impinging beam (E_i) intensity multiplied by (1) the cosine of the angle of incidence, and (2) the reflectivity (p) of the reflecting material and divided by (1) pi (π) and (2) the square of the distance, (d), from the reflecting surface to the eye. This relationship is expressed simply by this formula:

$$E_r = \frac{E_i \, (\cos \theta)(\rho)}{\pi \, d^2}$$

Consequently, it behooves a laser user to reduce the number and size of specular reflecting surfaces and, where possible, substitute diffuse reflecting materials. A myth should be dispelled here. It is falsely believed by some that by covering everything in a laser environment with black paint, a good diffuse reflector, there is less chance of a reflected beam causing harm. Two conditions result that actually could increase hazards. One is that the size of the pupil of the eye increases in darkness to allow more light to enter the ocular system thus increasing the potential eye damage from reflections in the beam path. The second condition, perhaps more important, is poor visibility in the room or laboratory, increasing the risk of accidents with electrical, chemical, or mechanical hazards. Laser environments should be well lighted except for beam detection or other short-timed purposes.

MANUFACTURED LASER REGULATIONS

The Radiation Control for Health and Safety Act of 1968 was passed by Congress to protect the public from the dangers of radiation from electronic products. The growing number of laser products in use in industry, research, and medicine has substantially increased the potential for radiation injuries, prompting the Food and Drug Administration to issue a performance standard for laser products, 21 CFR 1040.10 and 1040.11 under the Act.

A laser product is defined as any electronic product that consists of, incorporates, or is intended to incorporate a laser. A manufacturer is any person engaged in the business of manufacturing, assembling, or importing a laser product, whether or not he manufactures the laser. In addition, any person engaged in the business of modifying previously certified laser products in any way that affects the product's performance or intended function(s) pursuant to the standard is considered a manufacturer.

The performance standard was developed to protect public health and safety through the elimination or the containment of radiation that need not be accessible for the performance of the intended function of the laser product. For laser radiation that must be accessible, public health and safety is protected by providing for indication of the presence of laser radiation, by providing the user with certain means to control radiation, and by assuring adequate warnings, through use of product labels and instructions, to all personnel of the potential hazard. A fundamental concept in the performance standard is the classification of laser products based on the relative hazard of the level of accessible laser radiation. The standard establishes four classes of laser products based on the biological risk involved. Another fundamental concept in the standard is that the number of required safety performance features and the severity of the warning labels increases according to the classification assigned to the product. The standard also provides upper limits for accessible laser radiation from laser products manufactured for the specific purposes of surveying, leveling, alignment, or demonstration.

Federal regulations require that all laser products manufactured on or after August 2, 1976 be certified as complying with the performance standard. The manufacturer must demonstrate the product's compliance with the standard prior to certification or introduction into commerce by furnishing to the Bureau of Radiological Health reports pertaining to the radiation safety of the product and the associated quality control program. Annual reports must also be submitted summarizing the records required to be maintained. Failure to provide the required reports or product certification is a violation of Section 360B of the Radiation Control for Health and Safety Act of 1968 and can result in the imposition of civil penalties and restraint of dealers and distributors from selling or otherwise disposing of their laser products. Manufacturers of laser products who fail to comply with the performance standard are subject to corrective actions. Corrective actions may involve repurchase, repair, or replacement.

In July 1976 the National Institute for Occupational Safety and Health (NIOSH), Division of Biomedical and Behavioral Science, published a Laser Hazard Classification Guide, Interagency Agreement NOISH 1A-76-26, that had three objectives: (1) to gather technical specifications on all lasers currently in use from manufacturers' data sheets, (2) to classify the laser from the manufacturers' specifications, and (3) to confirm the manufacturers' data by measuring a small sample of lasers and cross-checking the calibration of the manufacturers' instrumentation. This report presents a computerized list of almost 2500 models available from over 175 manufacturers and distributors in the laser safety field.

The laser characteristics listed in this compendium for each manufacturer's model are: active medium, hazard class (based on ANSI Z136.1 definitions), wavelength (nm), diameter (cm), divergence (mr), beam shape (circular, rectangular, or other), output power/energy, pulse repetition rate (p/sec), pulse width (msec, μsec, or psec), irradiance (W/cm^2)/radiant exposure (J/cm^2), and comments on application.

If commercial laser users need information on the laser at hand or need advice on beam measurements, and other sources are not available, contact: Director (HFX-400), Division of Compliance, Bureau of Radiological Health, 5600 Fishers Lane, Rockville, Maryland 20857.

Noncommercial laser users must follow the same guidelines for hazard control as if the specially built laser were manufactured. It is especially important to label the beam outlet at the closest position to the chamber hole. This label should include the danger symbol, the laser output characteristics (such as CW or pulsed) and details of its maximum power or energy density, pulse width, and its ANSI Z136.1 class number.

REFERENCES

Hull, Daniel, M., Course 1, Introduction to Lasers, Laser/Electro Optics Technology Series, Center for Occupational Research and Development, Inc., Waco, Texas, 1982.

Heard, H. G., *Laser Parameter Measurements Handbook*, John Wiley and Sons, Inc., New York, New York, 1968.

Hecht, Jeff and Teresi, Dick, *Laser, Supertool of the 1980s*, Ticknor and Fields, New Haven, Connecticut, 1982.

2
Eye Components

It is not essential that laser users become experts on the biology and physiology of that very complicated organ, the eye. However, it is necessary to know how the laser beam of interest affects the functions of two of the eye's components: the *cornea* and the *retina*. In addition to being knowledgeable about how these two functional parts of the eye interact with various laser beam wavelengths, it is essential that the magnification factor of the complete eyeball assembly be understood. The ability of the eye to focus a certain wavelength range of radiation is one of the most fascinating and remarkable functions of any of the human body organs. Microscopes require extensive apparatus and lens systems to obtain magnifications that approach that capable of the human eye: 100,000 (10^5) times. This magnification factor is used in laser safety as an engineering constant in calculating optical density values.

The sole purpose for understanding the functions of these two eye components is to determine the selection of eye protection. This simple rule will help in this regard:

Laser wavelengths that are absorbed by the cornea do not reach the retina (by the high magnification factor of 10^5) and therefore do not require high optical density in the absorbing eyewear.

Not understanding this relationship of laser beam wavelength and absorption qualities of the cornea and retina can result in using unnecessarily expensive eyewear with poor visibility, as explained in Chapter 8. For example, clear (85-90% visual transmission) glass materials provide excellent protection for all ultraviolet and infrared wavelengths (above 0.90 μm). Colored plastics or glass materials are not needed in these ranges and indeed may be harmful with respect to not having good visibility in a laboratory environment.

EYE EXTERIOR: CORNEA

The chart in Figure 2.1 is presented for the purpose of understanding the role the cornea plays in evaluating damage of the eye from laser radiation. Note that the range of wavelengths from 0.35 to 1.4 μm is identified as the "ocular focus" region. All laser rays outside this region are absorbed by the outer components of the eye as shown graphically in Figure 2.2. There is some question about the corneal transmission of wavelengths as short as 0.35 μm; therefore, for the purpose of evaluating protective eyewear, as explained later, the absorption by the cornea of all wavelengths below 0.35 μm will be assumed.

EYE INTERIOR: RETINA

Evaluation of eye protection is so important in a laser safety program, that it is imperative that all laser personnel be aware of those lasers that emit rays in

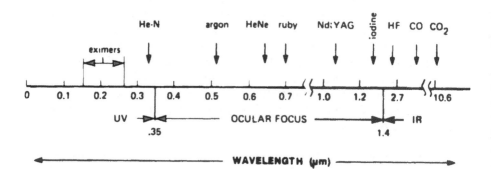

Figure 2.1 Location of the three regions of interest in evaluation of effect of laser light on eye components.

CORNEA OPTIC NERVE IRIS
LENS RETINA

ULTRAVIOLET
(<0.4 μm)

VISIBLE (FOCUSABLE)
(0.4-0.7 μm)

INVISIBLE (FOCUSABLE)
(0.7-1.4 μm)

INFRARED
(>1.4 μm)

Figure 2.2 Absorption and transmission of laser light by components of the ocular system.

the ocular focus region of the spectrum. Wavelengths from 0.35 to 1.40 μm are focused on the retina and are perceived by the eye as the following colors:

Color	Wavelength (μm)
Violet	0.35-0.42
Blue[a]	0.42-0.50
Green	0.50-0.54
Yellow	0.56-0.63
Red[a]	0.65-0.76
Invisible	0.76-1.40

[a]Primary colors.
Source: Encyclopaedia Britannica

As mentioned before, the transmission of the wavelengths of 0.35-0.40 μm to the retina is not understood as being as definite as the other colors; but in laser safety considerations it is practical to assume retinal sensitivity in this range—

it is safer. The same is true of the range 0.70-0.76 μm, which is out of the color band of wavelengths but is thought to be retinal sensitive. Therefore, it is recommended that, for practical application of risk factors, the region from 0.35 to 0.76 μm be considered the visible portion of the light spectrum.

In addition, as shown in Figure 2.2, wavelengths above 0.76 μm and extending to 1.40 μm are "invisible" to the eye but are focused on the retina by the ocular components. It is indeed strange that this noncolor range of wavelengths is focused on the retina for no apparent reason, but the same magnification factor of 10^5 applies and must be used in evaluating filters for eye protection. Fortunately, by another strange coincidence, clear glass materials are available to absorb these wavelengths (see Chap. 8).

LASER DAMAGE MECHANISMS

Biological damage details resulting from excessive laser absorption by eye components are of limited interest in a laser safety program. However, a short discussion may be helpful in understanding the effects of the various wavelengths and the modes of light propagation (continuous wave or pulsed).

Basically, a continuous wave (CW) laser causes damage by *thermal processes* that overheat the tissue. *Photochemical degeneration* also accompanies excessive CW laser radiation. The steady stream of photons is absorbed by tissue until the temperature rises above that of the cooling mechanism. In the case of the skin, the flowing blood temperature must be overcome by the absorption process until tissue cells are raised to excessive temperatures. In the eye, for those wavelengths outside the ocular focus region of 0.35-1.40 μm, the absorption occurs on the outer components, particularly the cornea which does not have a flowing blood system to cool its surface. Consequently, the damage threshold values of power density are lower for corneal damage than those for skin as revealed in Chapter 3.

The effect of thermal damage is displayed in the clinical use of argon or ruby radiation under controlled conditions to "spot weld" detached retinas. The temperature is adjusted by control of the laser beam input so that a joining of the adjacent tissues is accomplished without harming the deeper layers of the retinal wall. Also, when excessive laser radiation interacts with the skin, blisters form, as though from a burn; and charring occurs if additional energy is concentrated on a fixed position.

Pulsed lasers cause *blast damage* if the pulse duration is low. A pulsed laser is defined, for laser safety purposes, as having a pulse time or pulse-width of less than 0.25 sec; but, according to Arnold Goldman in his Ph.D. thesis*, the pulse durations have to be quite short to cause mechanical, or blast, damage: tens of nanoseconds or less. He states that, "Retinal lesions caused by these devices are the result of such stresses resulting from shock and acoustic waves generated at the site of energy deposition. In these lesions the pulse durations are so short that little or no thermal conduction occurs during the length of the pulse."

It was found in Dr. Goldman's experiments that retinal damage resulted from minute quantities of laser light concentrated on the retina as discussed in Chap. 3. These short bursts of light are analogous to a bullet. For example, a simple calculation shows that a one nanosecond pulse of light radiation is equivalent to approximately one foot in length:

$(3 \times 10^8 \text{ m/sec}) \times (1 \times 10^{-9} \text{ sec}) = 0.3 \text{ m}$

(speed of light) $\times$ (one nanosecond) $\cong$ one foot

Consequently, pulsed lasers are more hazardous to the eye, especially when the wavelength is in the ocular focus region.

---

*Arnold Goldman, An ultrastructural study of the effect of 30 ps pulses of 1064 nm laser radiation on the rhesus monkey retina, Virginia Commonwealth University, Richmond, Virginia, May 1975.

3

Laser-Eye Damage Thresholds

The key to determining the type of eye protection necessary is in understanding how the values of laser beam intensities that cause permanent damage are determined and what these values for the laser at hand are in terms of power density (W/cm^2) for continuous wave (CW) lasers and energy density (J/cm^2) for pulsed lasers. Fortunately, early in the evolution of laser technology, a keen interest in potential eye damage was shown by many researchers in this field. Animals were exposed to laser radiation of every available type and the damage evaluated to determine the intensity that caused the first indication of tissue degradation. Results were corroborated by independent teams of experimentalists and were used in determining the so-called "maximum permissible exposure" (MPE) limits in the standards for control of laser hazards, such as the American National Standard Institute's (ANSI) Standard No. Z136.1 "for the safe use of lasers" referred to in previous chapters.

The MPE values given in ANSI Z136.1 are determined by a committee that studies all available experimental information and arrives at a value somewhere below these known hazardous values. Tables are presented in the Standard that list MPE values for essentially all wavelengths from 0.220 to 10^3 μm and for exposure times from 10^{-9} to 3×10^4 sec. One table is for direct, or intrabeam, viewing; and another is for viewing a diffuse reflection. If an MPE value is desired from these tables, consideration must be given to such criteria as "limiting aperture," "extended sources," "multipulse limitations," and inconvenienced to the point of often being required to make calculations of various fractional exponential exponents of time, referred to graphs, and so on. It is often a complicated safari in a jungle of data.

Therefore, it appeared easier, and certainly more ethical, to use the raw data from experimentalists who determined the actual threshold damage information and use the lowest value found in the literature as the value of interest in calculating the optical density (O.D.) of filters to protect the eyes from specific wavelengths. If capable individuals wish to use the ANSI Z136.1 values of MPE, it should be pointed out that safety factors are used to reduce the experimental values, that these factors could be quite high. This could result, for example, in requiring darker eyewear than needed for adequate protection. Also, remember that these MPE values are to be used as guides in helping evaluate the laser environment.

The values shown in Tables 3.1 and 3.2 represent experimental data obtained from the literature from experiments with animal organs that closely approximate human eyes in structure and tissue, so that even these values are not precise. However, practical application of eye damage data is the concern of laser users, and an understanding of the method of securing these values is important. Then safety factors of known origin can be applied and understood. Historically, the CW information is well documented since several investigators

Table 3.1   Approximate damage threshold values for corneal tissue for various pulsed and CW lasers (minimum reported).

Available intensities	UV .2-.3 μm	IR 2.7 μm	10.6 μm
CW (W/cm^2)^b	10×10^{-3}	$(100 \times 10^{-3})^a$	100×10^{-3}
pulsed (J/cm^2):			
100 nsec	(4×10^{-3})	4×10^{-3}	(5×10^{-3})
30 nsec	(4×10^{-3})	(4×10^{-3})	(5×10^{-3})
20 nsec	(4×10^{-3})	(4×10^{-3})	(5×10^{-3})
1 nsec	(4×10^{-3})	4×10^{-3}	5×10^{-3}
30 psec	$(<4 \times 10^{-3})$	$(<4 \times 10^{-3})$	$(<5 \times 10^{-3})$

[a] All parenthetical values are estimates by the author.
[b] Assume one second maximum exposure.

Table 3.2 Approximate damage threshold values for retinal tissue for various pulsed and CW lasers (minimum reported).

Available intensities	Ocular focus	
	.5-.7 μm	1.1 μm
CW (W/cm^2)^b	10×10^{-3}	$<5 \times 10^{-3}$
pulsed (J/cm^2):		
100 nsec	$(100 \times 10^{-6})^a$	(100×10^{-6})
30 nsec	70×10^{-6}	(100×10^{-6})
20 nsec	130×10^{-6}	105×10^{-6}
1 nsec	(120×10^{-6})	(100×10^{-6})
30 psec	18×10^{-6}	9×10^{-6}

[a] All parenthetical values are estimates by the author.
[b] Assume one second maximum exposure.

undertook an intense study of the early lasers that were of the continuous beam type. Until recently it was more difficult to get pulsed laser data due to costs and availability of the laser systems. Most of the values listed for the fast-pulsed lasers in Tables 3.1 and 3.2 were obtained while the Los Alamos National Laboratory was assemblying large fusion experimental machines that had the small "front-end" assemblies well characterized and available for periods compatible with eye damage experimentors' schedules. In addition, that laboratory had an interest in supporting the experiments with laser and other facilities including animal care. Ironically, the values obtained from these valuable experiments were available but were not included in the ANSI Z136.1 1980 Standard because of lack of corroboration; it would appear however, that they could have been listed with asterisks and an explanation. Many cases of permanent damage to laser users' eyes in the past several years have resulted from exposure to ultrafast-pulse radiation. It does not take much energy density from these bullets of light to cause damage.

CORNEA

The cornea is the first exterior component of the human eye that would come in contact with laser radiation. As shown in Figure 3.1, transmission of light

Figure 3.1 Transmission curve for human ocular system of wavelengths in the ocular focus region.

radiation is permitted through the cornea for the region of wavelengths from approximately 0.35 to 1.40 μm. Not all of these rays penetrate to the retina, of course; but, as the curve shows, a high percentage of the visible rays (0.4-0.7 μm) do so. A lower percentage of the invisible portion (0.7-1.4 μm) is transmitted. All other wavelengths are essentially 100% absorbable by the cornea. The minimum amount of laser radiation that can cause permanent damage is defined as the threshold value. Damage threshold values are shown in units of watts per square centimeter (W/cm^2) for one-second exposures of CW laser radiation, and in units of joules per square centimeter (J/cm^2) for pulsed laser radiation. Pulsed laser data also includes pulse duration because the shorter the pulse width, the lower the value for the same wavelength. These data were obtained from experiments with rabbit corneas which closely resemble human corneas.

The use of these damage threshold values will be described in Chapter 8 when we discuss selecting protective eyewear.

RETINA

It is quite difficult to obtain data from experiments that involve destruction of a complete eyeball of an animal to recover retinal tissue in measuring damage thresholds. However, the selection of the rhesus monkey as the sacrificial animal is fortuitous because of its availability and because its eye closely resembles the human eye in receiving light waves. The values listed in Table 3.2 for retinal damage thresholds were obtained by a tedious biological procedure from monkey eyes that will not be described here but which is available from the references written by principal investigator William Ham. Use of these data in calculating optical density of protective eyewear will be discussed in Chapter 8.

If retinal lesions are not permanent and vision is restored, then the radiation has not reached damage threshold. This probably happens more frequently than reported as a result of accidents with relatively low-power pulsed lasers by operators not wearing eyewear and guessing that the beam intensity is too low to cause damage. This philosophy permeates the personal testimony in several documented cases reported by Boldrey (1981). No permanent damage has been reported by laser users wearing laser protective eyewear properly.

The following quotation from the editorial page "Comment" of the August 1977 edition of *Laser Focus* describes one laser accident victim's view. It is included here to illustrate the result of discounting or not heeding the effect of very "low-power" pulsed lasers.

> The necessity for safety precautions with highpower lasers was forcibly brought home to me last January when I was partially blinded by a reflection from a relatively weak neodymium-yag laserbeam. Retinal damage resulted from a 6-millijoule, 10-nanosecond pulse of invisible 1064-nanometer radiation. I was not wearing protective goggles at the time, although they were available in the laboratory. As any experienced laser researcher knows, goggles not only cause tunnel vision and become fogged, they become very uncomfortable after several hours in the laboratory.
>
> When the beam struck my eye I heard a distinct popping sound, caused by a laser-induced explosion at the back of my eyeball. My vision was obscured almost immediately by streams of blood floating in the vitreous humor, and by what appeared to be particulate matter suspended in the vitreous humor. It was like viewing the world through a round fishbowl full of glycerol into which a quart of blood and a handful of black pepper have been partially mixed. There was local pain within a few minutes of the accident, but it did not become excruciating. The most immediate response after such an accident is horror. As a Vietnam War Veteran, I have seen several terrible scenes of human carnage, but none affected me more than viewing the world through my bloodfilled eyeball. In the aftermath of the accident I went into shock, as is typical in personal injury accidents.

As it turns out, my injury was severe but not nearly as bad as it might have been. I was not looking directly at the prism from which the beam had reflected, so the retinal damage is not in the fovea. The beam struck my retina between the fovea and the optic nerve, missing the optic nerve by about three millimeters. Had the focused beam struck the fovea, I would have sustained a blind spot in the center of my field of vision.

Figure 3.2 Diagram showing results of the retinal examination of victim of laser radiation accident. Eye damage caused by laser pulse is shown in this plot of field of view under high-intensity illumination solid lines. Outer circles show field of view; the two small regions inside the field of view are blind spots produced by laser damage. The blind spots are larger than the lesion and occupy a larger area under low illumination.

Had it struck the optic nerve, I probably would have lost the sight of that eye.

The beam did strike so close to the optic nerve, however, that it severed nerve-fiber bundles radiating from the optic nerve. This has resulted in a cresent-shaped blind spot many times the size of the lesion. The diagram (Figure 3.2) is a Goldman-Fields scan of the damaged eye, indicating the sightless portions of my field of view four months after the accident. The small blind spot at the top exists for no discernible reason; the lateral blind spot is the optic nerve blind spot. The effect of the large blind area is much like having a finger placed over one's field of vision. Also I still have numerous floating objects in the field of view of my damaged eye, although the blood streamers have disappeared. These "floaters" are more a daily hinderance than the blind areas, because the brain tries to integrate out the blind area when the undamaged eye is open. There is also recurrent pain in the eye, especially when I have been reading too long or when I get tired.

The moral of all this is to be careful and to wear protective goggles when using highpower lasers. The temporary discomfort is far less than the permanent discomfort of eye damage. The type of reflected beam which injured me also is produced by the polarizers used in q switches, by intracavity diffraction gratings, and by all beamsplitters or polarizers used in optical chains.

Laser-induced retinal damage suffered by those in laser environments is relatively rare, but, because of the increased use of this magnificent tool, more and more inexperienced people are getting involved with lasers, and uninformed users could be involved in accidental excessive exposure. Notice in the final paragraph of the accident victim's testimony that the 6-millijoule (6×10^{-3} J), 10-nanosecond (10×10^{-9} s) pulse of neodymium-yag radiation (1.06 μm) is referred to as "highpower." Also, eyewear need not be undesirable as described in the first paragraph, "goggles not only cause tunnel vision and become fogged, they become very uncomfortable after several hours in the laboratory." Lightweight, comfortable laser eyewear will be discussed in Chapter 8.

REFERENCES

Egbert, D. E., and Maker, E. F. (December 1977). Corneal damage thresholds for infrared laser exposure: Empirical data, model predictions, and safety standards. Final Report for Period April 1975 to June 1977, USAF School of Aerospace Medicine, Brooks Air Force Base, Texas.

Hamm, W. T., Jr., Mueller, H. A., and Goldman, A. I. (August 1975). Threshold data on retinal lesions in the rhesus monkey from picosecond pulses of 1064 nm and 532 nm laser radiation. A report to the Los Alamos Scientific Laboratory, L-Division.

4

Skin Damage Thresholds

Skin damage from laser radiation is not as great a concern as eye damage; such skin injury can be treated similarly to treatment of a thermal burn or wound. Further, in power or energy densities high enough to cause skin damage, the laser beam is usually enclosed, or some form of physical protection is provided for personnel. If this is not done, laser operators must wear protective clothing. Knowledge of skin damage thresholds also helps to evaluate optical densities of eyewear; if the need for high optical densities (above 5 or 6 in most cases) is specified, simple calculations will show that the power or energy density anticipated for the laser beam will be above the skin damage threshold as explained in specific cases in Chapter 8.

Table 4.1 presents data for skin damage obtained by pioneers in the field of laser safety. Dr. Leon Goldman, a dermatologist, had a clinical interest in the effect of laser radiation on skin tissue, and his associate, James Rockwell, pioneered the physics of lasers adapted to medical and biological applications. Some of their publications are listed in the references at the end of this chapter.

Experiments were conducted at the Los Alamos National Laboratory to obtain the fast-pulsed laser damage threshold data for hydrogen-fluoride (HF) and carbon dioxide (CO_2) wavelengths. Piglets were used because their skin (Fig. 4.1) closely resembles Caucasian human skin, and therefore results of laser beam absorption are accepted as equivalent. Damage values for darker

Table 4.1 Skin damage threshold values for laser radiation (Caucasian).

| | UV | Ocular focus | | IR | |
	$.2-.3\ \mu m$	$.5-.7\ \mu m$	$1.1\ \mu m$	$2.7\ \mu m$	$10.6\ \mu m$
Available intensities					
CW (W/cm^2)[b]	3×10^{-3}	4	28	(3)[a]	3
pulsed (J/cm^2)[b]:					
100 nsec	(300×10^{-3})	110×10^{-3}	2.2	300×10^{-3}	(300×10^{-3})
30 nsec	(200×10^{-3})	(100×10^{-3})	(2.0)	(200×10^{-3})	(200×10^{-3})
20 nsec	(200×10^{-3})	(100×10^{-3})	(2.0)	(200×10^{-3})	(200×10^{-3})
1 nsec	(200×10^{-3})	(100×10^{-3})	(2.0)	(200×10^{-3})	230×10^{-3}
30 psec	(200×10^{-3})	(100×10^{-3})	(2.0)	(200×10^{-3})	(200×10^{-3})

[a] All parenthetical values are estimates by the author.
[b] Assume one second maximum exposure.

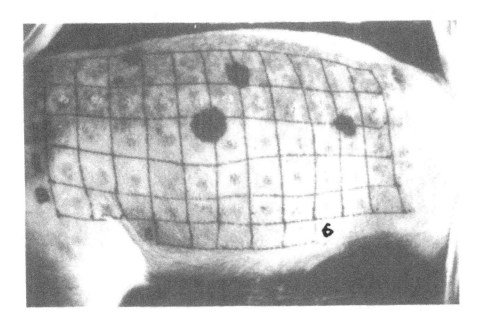

Figure 4.1 Photograph of piglet skin used in damage threshold experiments. (Courtesy of James Rockwell and Los Alamos National Laboratory.)

human skin are somewhat lower, but only Caucasian skin values will be presented in the Los Alamos results with fast-pulsed lasers.

Skin damage threshold values for laser radiation are close to those of glass, plastic, and other non-tissue materials, as described in government studies listed in the references at the end of this chapter. This is an important point, eyewear lenses and frames are subject to damage if laser beams capable of skin damage are not enclosed, even if protective clothing is used. Incineration of combustible materials is also possible above these values of skin damage. There should be no plausible reason for personnel to be exposed to the risk of laser radiation above that which could cause skin damage.

REFERENCES

Goldman, L., and Rockwell, R. J., Jr. (1971). *Lasers in Medicine*, Gordon and Breach, New York.

Rockwell, R. J., Jr., and Goldman, L. (June 1974). Research on human skin laser damage threshold. Final Report, Contract F41609-72-C-0007, USAF School of Aerospace Medicine, Brooks Air Force Base, Texas, prepared by Department of Dermatology and Laser Lab Medical Center, University of Cincinnati.

Rockwell, R. J., Jr. (March 1976). Damage thresholds on pig skin following irradiation by short pulse hydrogen flouride and carbon dioxide lasers. Report Purchase Order No. LP5-91068-1, Los Alamos Scientific Laboratory.

Envall, K. R., and Murray, Ross, Jr. (May 1979). Evaluation of commercially available laser protective eyewear. HEW Publications (FDA) 79-8086, BRH, Rockville, Maryland 20857.

Stevison, D. F., and Olson, J. C. (1977). Damage survey of safety glasses irradiated with CO_2 laser energy. Report No. AFML-TR-77-186, Air Force Materials Laboratory, Wright-Patterson Air Force Base, Ohio 45433.

5

American National Standards Institute's ANSI Z136.1 "Safe Use of Lasers"

All laser users in the United States will soon probably have to comply with the basic tenets of the American National Standards Institute's Standard Number ANSI Z136.1. This standard was developed by the first committee Z136 in the late 1960s. The committee issued the first document in 1973 with a revised edition in 1976. A revision of the 1976 document was made by the committee in 1980, and the latest edition, dated 1986, is expected to be promulgated into law by Congress. Several states are currently using the standard as a guide for interim state laws.

A brief description of the American National Standards Institute is presented here to familiarize laser users with this unique nonprofit organization and to explain why standards approved by this body are models for federal, state, and even international regulations.

The ANSI Z136.1-1986 is one of more than 8500 standards approved to date by the American National Standards Institute.

The Standards Institute provides the machinery for creating voluntary standards. It serves to eliminate duplication of standards activities and to weld conflicting standards into single, nationally accepted standards under the designation "American National Standards."

Each standard represents general agreement among manufacturers, sellers, and user groups as to the best current practice with regard to some specific problem. Thus, the completed standards cut across the whole fabric of production, distribution, and consumption of goods and services. American National Standards, by reason of Institute procedures, reflect a national consensus of manufacturers, consumers, and scientific, technical, and professional organizations, and government agencies. The completed standards are used widely by industry and commerce and often by municipal, state, and federal governments.

The Standards Institute, under whose auspices this work is being done, is the U.S. clearinghouse and coordinating body for voluntary standards activity on the national level. It is a federation of trade associations, technical societies, professional groups, and consumer organizations. Some 1000 member companies are affiliated with the Institute.

The American National Standards Institute is the United States' representative of the International Organization for Standardization (ISO) and the International Electrotechnical Commission (IEC). Through these channels, U.S. standards interests make their positions felt on the international level. American National Standards are on file in the libraries of the national standards bodies of more than 60 countries. The American National Standards Institute, Inc. is located at 1430 Broadway, New York, New York 10018.

The ANSI Z136.1-1986 document is a nongovernment publication. It was adopted on October 23, 1981, and is approved for use by the Department of the Navy, the Department of Defense, and other federal agencies. Contractors of these agencies are required to consider this standard as regulatory.

An American National Standard implies a consensus of those substantially concerned with its scope and provisions. An American National Standard is intended as a guide to aid the manufacturer, the consumer, and the general public. The existence of an American National Standard does not in any respect preclude anyone, whether he has approved the standard or not, from manufacturing, marketing, purchasing, or using products, processes, or procedures not conforming to the standard. American National Standards are subject to periodic review and users are cautioned to obtain the latest editions.

An American National Standard may be revised or withdrawn at any time. The procedures of the American National Standards Institute require that action be taken to reaffirm, revise, or withdraw a standard no later than five years from the date of publication. Purchasers of American National Standards may receive current information on all standards by calling or writing the American National Standards Institute.

The foreword in the laser user document is quoted as follows.*

---

*The quotations from ANSI-Z136.1 material are reproduced with permission from American National Standard Institute's publisher, the Laser Institute of America. Copies of this standard may be purchased from the Laser Institute of America, 12424 Reserach Parkway, Suite 130, Orlando, FL 32826.

This American National Standard provides guidance for the safe use of lasers and laser systems by defining control measures for each of four laser classifications. Once a laser or laser system is properly classified, there should be no need to carry out tedious measurements or calculations in order to meet the provisions of the standard. However, technical information on measurements, calculations, and biological effects is also provided within the standard and its appendixes.

This standard is a revision of American National Standard Z136.1-1980. It is expected that this standard will continue to be periodically revised as new information and experience in the use of lasers are gained.

... Suggestions for improvement of this standard will be welcome. They should be sent to the Laser Institute of America. Copies of this standard may be ordered from this address with price quotations included.

This standard was processed and approved for submittal to ANSI by American National Standards Committee Z136 on the "Safe Use of Lasers," whose scope covers protection against hazards associated with the use of lasers and optically radiating diodes. Committee approval of the standard does not necessarily imply that all members voted for its approval. At the time it approved this standard, the Z136 Committee had a total of 51 representatives or alternates from 33 organizations and 15 individual members.

Eleven subcommittees participated in the development of the Safe Use of Lasers Standard. They included subcommittees on

1. Biological effects of lasers on the eye
2. Biological effects of lasers on the skin
3. Measurements and hazard evaluation of laser exposure
4. Laser control measures
5. Elements of a laser safety and training program
6. Medical surveillance for laser exposure
7. Special considerations associated with laser study
8. Terminology
9. Editorial
10. Fiber optics
11. Laser safety in medical facilities

The author has served on the Elements of a Laser Safety and Training Program subcommittee and is currently serving on the Laser Control Measures subcommittee as an individual (unlisted) member.

The scope, objectives, and philosophy of the standard are contained in Section 1.

1. General

1.1 Scope. This standard provides reasonable and adequate guidance for the safe use of lasers and laser systems.

1.2 Application. The objective of this standard is to provide reasonable and adequate guidance for the safe use of lasers and laser systems. A practical means for accomplishing this is first to classify lasers and laser systems according to their relative hazards and then to specify appropriate controls for each classification.

The basis of the hazard classification scheme in Section 3 of this standard is the ability of the primary laser beam or reflected primary laser beam to cause biological damage to the eye or skin during intended use. For example, a Class 1 laser is one that is considered to be incapable of producing damaging radiation levels and is, therefore, exempt from any control measures or other forms of surveillance. Class 2 lasers (low-power) are divided into two subclasses, 2 and 2a. A Class 2 laser emits in the visible portion of the spectrum (0.4 to 0.7 μm) and eye protection is normally afforded by the aversion response including the blink reflex. Class 3 lasers (medium-power) are divided into two subclasses, 3a and 3b. A Class 3 laser may be hazardous under direct and specular reflection viewing conditions, but the diffuse reflection is usually not a hazard. A Class 3 laser is normally not a fire hazard. A Class 4 laser (high-power) is a hazard to the eye and skin from the direct beam and sometimes from a diffuse reflection and also can be a fire hazard.

It must be recognized that the classification scheme given in this standard relates specifically to the laser product and its potential hazard, based on operating characteristics. However, the conditions under which the laser is used, the level of safety training of individuals using the laser and other environmental and personnel factors are important considerations in determining the full extent of safety control measures. Since such situations require informed judgments by responsible persons, major responsibility for such judgments has been assigned to a person with the requisite authority and responsibility, namely the Laser Safety Officer (LSO).

Lasers or laser systems certified for a specific class by a manufacturer in accordance with the Federal Laser Product Performance Standard may be considered as fulfilling all classification requirements of this standard.

In cases where the laser or laser system classification is not provided or where the class level may change because of the addition or deletion of engineering control measures (see 4.3), the laser or laser system shall be classified by the LSO in accordance with the descriptions given in Section 3 or the methods described in Section 9, or both.

The recommended stepwise procedure for using this standard is as follows:

1. Determine the appropriate class of laser or laser system.
2. Comply with the measures specified for that class of laser or laser system, using the following table as a guide. This procedure will in most cases eliminate the need for measurement of laser radiation, quantitative analysis of hazard potential, or use of the Maximum Permissible Exposure (MPE) values given in Section 8 of this standard.

Class	Control Measures	Medical Surveillance
1	Not applicable*	Not applicable
2	Applicable	Not applicable
2a	Applicable	Not applicable
3a	Applicable	Not applicable
3b	Applicable	Applicable
4	Applicable	Applicable

*During normal operation and maintenance only.
Alignment and service procedures of an embedded Class 2, 3, or 4 laser shall require control or administrative procedures appropriate to the class during these functions.

For the purposes of this standard, except for short-distance viewing of small diameter focused Class 3 lasers (see 3.4.1.3), only Class 4 lasers are capable of producing hazardous diffuse reflections, hence calculations for viewing diffuse reflections from Class 1, 2, and 3 systems are normally unnecessary.

The laser hazard classification system is based entirely on the laser radiation emission. Other ancillary hazards must be dealt with separately and are addressed in Section 7 (Special Considerations).

Other special application standards within the Z136 series may have provisions which are exceptions to the requirements of this standard.

Each exception is valid only for applications within the scope of the standard in which it appears.

1.3 Laser Safety Officer (LSO)

1.3.1 General. An individual shall be designated the Laser Safety Officer (LSO) with the authority and responsibility to monitor and enforce the control of laser hazards and to effect the knowledgeable evaluation and control of laser hazards.

Depending on the extent and number of laser installations, the position of LSO may or may not be a full time assignment. In some instances, designation of an LSO may not be required. Such instances could include operation and maintenance which are limited to Class 1 and Class 2 lasers. However, under some circumstances it may be necessary to designate an LSO for example, if service is performed on a laser product having an embedded Class 3a, Class 3b, or Class 4 laser. In such instances, the designation of an LSO may be the responsibility of the organization requiring access to the embedded laser or laser system, such as the service company or organization.

There shall be a designated LSO for all circumstances of operation, maintenance, and service of a Class 3a, Class 3b or Class 4 laser or laser system.

1.3.2 LSO Specific Responsibilities

1.3.2.1 Classification. The LSO shall classify, or verify classifications, of lasers and laser systems used under the LSO's jurisdiction.

1.3.2.2 Hazard Evaluation. The LSO shall be responsible for hazard evaluation of laser work areas, including the establishment of Nominal Hazard Zones (NHZ). (See 3.4.)

1.3.2.3 Control Measures. The LSO shall be responsible for assuring that the prescribed control measures are in effect, recommending or approving substitute or alternate control measures when the primary ones are not feasible or practical, and periodically auditing the functionability of those controls measures in use.

1.3.2.4 Procedure Approvals. The LSO shall approve standard operating procedures, alignment procedures, and other procedures that may be part of the requirements for administrative and procedural control measures.

1.3.2.5 Protective Equipment. The LSO shall recommend or approve protective equipment i.e., eyewear, clothing, barriers, screens, etc., as may be required to assure personnel safety. The LSO shall assure that protective equipment is audited periodically to ensure proper working order.

1.3.2.6 Signs and Labels. The LSO shall approve the wording on area signs and equipment labels.

1.3.2.7 Facility and Equipment. The LSO shall approve laser installation facilities and laser equipment prior to use. This also applies to modification of existing facilities or equipment.

1.3.2.8 Training. The LSO shall assure that adequate safety education and training is provided to laser area personnel.

1.3.2.9 Medical Surveillance. The LSO shall determine the personnel categories for medical surveillance. (See 6.2.)

1.3.2.10 Additional. Additional recommended duties of the LSO are included in the Appendix.

6

Classification of Lasers by ANSI Z136.1

As stated in Section 1.2 of the ANSI Z136.1 Standard*, "The basis of the hazard classification scheme in this standard is the ability of the primary laser beam or reflected primary beam to cause biological damage to the eye or skin." Section 2 of the standard is "Definitions," and will not be discussed here. However, it is important that the classification of lasers be understood by laser users and the following section will explain the scheme.

3. Hazard Evaluation and Classification

3.1 General. Three aspects of a laser application influence the total hazard evaluation and thereby influence the application of control measures: (1) the laser or laser system's capability of injuring personnel, (2) the environment in which the laser is used, and (3) the personnel who may use or be exposed to laser radiation.

The laser classification scheme is based on aspect 1. Laser and laser systems classified in accordance with this standard shall be labeled with the appropriate hazard classification (see 3.3). Where appropriate, the

*Published by the Laser Institute of America, 12424 Research Parkway, Suite 130, Orlando, FL 32826.

classification labeling from the Federal Laser Products Performance Standards may be used to satisfy this labeling requirement. It should be noted that in some cases there may be differences between this standard and the Federal Laser Product Performance Standard. (See Appendix C for references.) If the laser has been modified subsequent to classification by the manufacturer, the guidance in 4.10 shall be used.

Any laser or laser system shall be classified according to its accessible radiation during normal operation and maintenance.

Aspects (2) and (3) vary with each laser application and cannot be readily standardized. The total hazard evaluation procedure shall consider all three aspects, although in most cases only aspect (1) influences the control measures which are applicable.

The detailed hazard evaluation or the classification of a laser or laser system as outlined in this section should not be attempted by personnel untrained in laser safety or optical engineering/physics. When it is necessary for the LSO to effect such an evaluation, it should be based on adequate training or knowledgeable assistance so that the calculations and risk determinations outlined in this section will be accomplished. Errors in such analyses could result in injuries of personnel and should not be taken lightly.

3.2 Laser Considerations. The LSO or knowledgeable individual responsible for laser classification shall ensure that laser-output data are valid in accordance with Section 9 (Measurements). Classification shall be based on the maximum output available for the intended use.

3.2.1 Multiwavelength Lasers. The classification of lasers or laser systems capable of emitting numerous wavelengths shall be based on the most hazardous possible operation.

3.2.1.1 A multiwavelength laser which by design can operate only as a single-wavelength laser shall be classified as a single-wavelength laser.

3.2.1.2 A multiwavelength laser which by design can operate over two or more wavelength regions (as defined in Section 8) shall require classification in each region of operation. The appropriate control measures for each region shall be taken.

3.2.2 Repetitively Pulsed Lasers. The evaluation of repetitively pulsed lasers or exposures requires the use of certain correction factors.

3.2.3 Radiometric Parameters of the Laser Required for Determining Laser Hazard Classification.

Table 6.1 Accessible Emission Limits for Selected Continuous-Wave[a] Lasers and Laser Systems

Wavelength range (μm)	Emission duration (s)	Class 1[b]	Class 2[c]	Class 3[d]	Class 4
Ultraviolet 0.2 to 0.4	3×10^4	$\leq 0.8 \times 10^{-9}$ W to $\leq 8 \times 10^{-6}$ W depending on wavelength (see Table 5)	—	> Class 1 but ≤ 0.5 W depending on wavelength (see Table 5)	>0.5 W
Visible 0.4 to 0.7	3×10^4	$\leq 0.4 \, c_B \times 10^{-6}$ W	> Class 1 but $\leq 1 \times 10^{-3}$ W	> Class 2 but ≤ 0.5 W	>0.5 W
Near Infrared 0.7 to 1.06	3×10^4	$\leq 0.4 \times 10^{-6}$ W to $\leq 200 \times 10^{-6}$ W depending on wavelength (see Fig. 9)	—	> Class 1 but ≤ 0.5 W depending on wavelength (see Fig. 9)	>0.5 W
Near Infrared 1.06 to 1.4	3×10^4	$\leq 200 \times 10^{-6}$ W	—	> Class 1 but ≤ 0.5 W	>0.5 W
Far Infrared 1.4 to 10^2	>10	$\leq 0.8 \times 10^{-3}$ W	—	> Class 1 but ≤ 0.5 W	>0.5 W
Submillimeter 10^2 to 10^3	>10	≤ 0.1 W	—	> Class 1 but ≤ 0.5 W	>0.5 W

[a] Emission duration >0.25 s.

[b] When the design or intended use of the laser or laser system ensures personnel exposures of less than 10^4 s in any 24-hour period, the limiting exposure duration may establish a higher exempt power level, as discussed in 3.2.3.

[c] See 3.3.2.2 for explanation of Class 2a laser.

[d] For 1 to 5 mW cw laser systems (Class 3a) see 3.3.3.1 and 3.3.3.2.

3.2.3.1 Classification of essentially all lasers requires the following parameters:

1. Wavelength(s) or wavelength range
2. For cw or repetitively pulsed lasers: average power output and limiting exposure duration τ_{max} in a period of 3×10^4 s inherent in the design or intended use of the laser or laser system.
3. For pulsed lasers: total energy per pulse (or peak power), pulse duration, prf, and emergent beam radiant exposure.

3.3 Laser and Laser System Hazard Classification Definitions. Tables 6.1 and 6.2 offer a summary of levels for laser and laser system classification: Table 6.1 for cw lasers with an emission duration $\geqslant 0.25$ s and Table 6.2 for pulsed lasers with an emission duration < 0.25 s.

3.3.1 Class 1 Lasers and Laser Systems

3.3.1.1 Any laser, or laser system containing a laser, that cannot emit accessible laser radiation levels in excess of the Class 1 AEL for the maximum possible duration inherent in the design or intended use of the laser or laser system is a Class 1 laser or laser system during normal operation and is exempt from all control measures or other forms of surveillance with the exception of applicable requirements for embedded lasers. The exemption strictly applies to emitted laser radiation hazards and not to other potential hazards.

3.3.2 Class 2 and Class 2a Visible Lasers and Laser Systems

3.3.2.1 Class 2 lasers and laser systems include:

1. Visible (0.4 to 0.7 μm) cw lasers and laser systems which can emit accessible radiant power exceeding the Class 1 AEL for the maximum possible duration inherent in the design or intended use of the laser or laser system, but not exceeding 1 mW
2. Visible (0.4 to 0.7 μm) repetitively pulsed lasers and laser systems which can emit accessible radiant power exceeding the appropriate Class 1 AEL for the maximum possible duration inherent in the design or intended use of the laser or laser system, but not exceeding the Class 1 AEL for a 0.25 s exposure duration.

3.3.2.2 Visible (0.4 to 0.7 μm) laser and laser systems intended for a specific use where the output is not intended to be viewed shall be designated Class 2a by the LSO, provided that the accessible radiation does

Table 6.2 Summary of Levels (Energy and Radiant Exposure Emissions) for Single-Pulsed Laser and Laser System Classification[a]

Wavelength range (μm)	Emission duration (s)	Class 1	Class 3	Class 4
Ultraviolet 0.2 to 0.4	$>10^{-2}$	$\leq 24 \times 10^{-6}$ J to 7.9×10^{-3} J	$>$ Class 1 but ≤ 10 J·cm^{-2}	>10 J·cm^{-2}
Visible 0.4 to 0.7	10^{-9} to 0.25	$\leq 0.2 \times 10^{-6}$ J	$>$ Class 1 but $\leq 31 \times 10^{-3}$ J·cm^{-2}	$>31 \times 10^{-3}$ J·cm^{-2}
Near Infrared[b] 0.7 to 1.06	10^{-9} to 0.25	$\leq 0.25 \times 10^{-3}$ J	$>$ Class 1 but $\leq 10 \cdot$ f cm^{-2}	>10 J·cm^{-2}
		$\leq 0.2 \times 10^{-6}$ to 2×10^{-6} J	$>$ Class 1 but $\leq 31 \times 10^{-3}$ J·cm^{-2}	$>31 \times 10^{-3}$ J·cm^{-2}
1.06 to 1.4	10^{-9} to 0.25	$\leq 0.25 \times 10^{-3}$ to 1.25×10^{-3} J	$>$ Class 1 but ≤ 10 J·cm^{-2}	>10 J·cm^{-2}
		$\leq 2 \times 10^{-6}$ J	$>$ Class 1 but $\leq 31 \times 10^{-3}$ J·cm^{-2}	$>31 \times 10^{-3}$ J·cm^{-2}
Far Infrared 1.4 to 10^{2}	10^{-9} to 0.25	$\leq 1.25 \times 10^{-3}$ J	$>$ Class 1 but ≤ 10 J·cm^{-2}	>10 J·cm^{-2}
		$\leq 80 \times 10^{-6}$ J	$>$ Class 1 but ≤ 10 J·cm^{-2}	>10 J·cm^{-2}
Submillimeter 10^{2} to 10^{3}	10^{-9} to 0.25	$\leq 3.2 \times 10^{-3}$ J	$>$ Class 1 but ≤ 10 J·cm^{-2}	>10 J·cm^{-2}
		$\leq 10 \times 10^{-3}$ J	$>$ Class 1 but ≤ 10 J·cm^{-2}	>10 J·cm^{-2}
	to 0.25	≤ 0.4 J	$>$ Class 1 but ≤ 10 J·cm^{-2}	>10 J·cm^{-2}

[a] There are no Class 2 single-pulsed lasers.

[b] Diffuse reflection criteria apply from 10^{-9} to 33×10^{-3} s for Class 3. For $>33 \times 10^{-3}$ s exposure, the maximum radiant exposure is 10 J · cm^{-2}. Class 1 and 3 values are wavelength dependent (see Fig. 8).

not exceed the Class 1 AEL for an exposure duration less than or equal to 10^3 s.

3.3.3 Class 3a and Class 3b Lasers and Laser Systems

3.3.3.1   Class 3a and Class 3b laser and laser systems include:

1. Infrared (1.4 μm to 1 mm) and ultraviolet (0.2 to 0.4 μm) lasers and laser systems which can emit accessible radiant power in excess of the Class 1 AEL for the maximum possible duration inherent in the design of the laser or laser system, but that (a) cannot emit an average radiant power in excess of 0.5 W for $\geqslant$0.25 s or (b) cannot produce a radiant exposure of 10 J$\bullet$cm^{-2} within an exposure time <0.25 s

2. Visible (0.4 to 0.7 μm) cw or repetitively pulsed lasers and laser systems that produce accessible radiant power in excess of the Class 1 AEL for a 0.25 s exposure time (1 mW for a cw laser), but that cannot emit an average radiant power greater than 0.5 W

3. Visible and near-infrared (0.4 to 1.4 μm) single-pulsed lasers that can emit accessible radiant energy in excess of the Class 1 AEL but which cannot produce a radiant exposure that exceeds 10 J$\bullet$cm^{-2} or produce a hazardous diffuse reflection

4. Near-infrared (0.7 to 1.4 μm) cw lasers or pulsed lasers which can emit accessible radiant power in excess of the Class 1 AEL for the τ_{max} inherent in the design or intended use of the laser or laser system, but which cannot emit an average power of 0.5 W or greater for periods $\geqslant$0.25 s

3.3.3.2   All Class 3 lasers and laser systems which have an accessible output power between 1 and 5 times the Class 1 AELs for wavelengths less than 0.4 μm or greater than 0.7 μm, or the Class 2 AELs for wavelengths between 0.4 μm and 0.7 μm, are Class 3a.

3.3.3.3   All Class 3 lasers and laser systems which do not meet the requirements of 3.3.3.2 shall be classified as Class 3b.

3.3.4   **Class 4 Lasers and Laser Systems.**   Class 4 lasers and laser systems include:

1. Ultraviolet (0.2 to 0.4 μm) and infrared (1.4 μm to 1 mm) lasers and laser systems which (a) emit an average accessible radiant power in excess of 0.5 W for periods $\geqslant$0.25 s or (b) produce a

radiant exposure of 10 J • cm^{-2} for an exposure duration of <0.25 s

2. Visible (0.4 to 0.7 μm) and near-infrared (0.7 to 1.4 μm) lasers and laser systems which (a) emit an average accessible radiant power of 0.5 W or greater for periods $\geqslant$0.25 s or (b) produce a radiant exposure in excess of 10 J • cm^{-2}, or a hazardous diffuse reflection for periods <0.25 s

3.4.1 Indoor Laser Operations. In general, only the laser is considered in evaluating an indoor laser operation if the beam is enclosed or is operated in a controlled area.

3.4.2 Outdoor Laser Operations Over Extended Distances. The total hazard evaluation of a particular laser system depends on defining the extent of several potentially hazardous conditions. In this evaluation, consider all optics (lenses, mirrors, fiber optics etc.) which are a permanent part of the laser beam path.

3.5 Personnel. The personnel who may be in the vicinity of a laser and its emitted beam(s), including maintenance personnel, service personnel, and the operator, can influence the total hazard evaluation and hence influence the decision to adopt additional control measures not specifically required for the class of laser being employed. This depends upon the classification of the laser or laser system.

3.5.1 If children or others unable to read or understand warning labels may be exposed to potentially hazardous laser radiation, the evaluation of the hazard is affected and control measures may require appropriate modification.

3.5.2 The type of personnel influences the total hazard evaluation. It must be kept in mind that for certain lasers or laser systems (for example, military laser rangefinders and some lasers used in the construction industry), the principal hazard control rests with the operator, whose responsibility is to avoid aiming the laser at personnel or flat mirrorlike surfaces.

The following are considerations regarding operating personnel and those who may be exposed:

1. Maturity of judgment of the laser user(s)
2. General level of training and experience of laser user(s) (that is, whether high school students, military personnel, production line operators, scientists, etc.)

3. Awareness on the part of onlookers that potentially hazardous laser radiation may be present and the relevant safety precautions

4. Degree of training in laser safety of all individuals involved in the laser's operation

5. Reliability of individuals to follow standard operating procedures (SOP) and recommended control procedures

6. Number of individuals and their location relative to the primary beam or reflections, and potential for accidental exposure

7. Other hazards not due to laser radiation which may cause the individuals to react unexpectedly or which influence the choice of personnel protective equipment.

7

Engineering Control Measures
of ANSI Z136.1

Engineering controls are the fundamental tenets of a sound laser safety program. Because the ANSI Z136.1 standard has detailed the requirements for a safe laser environment through explicit control measures, portions of Section 4 are presented here.* The verb *shall* in the standard denotes a requirement that is mandatory, whereas the verb *may* denotes a recommendation.

4. Control Measures

4.1 General Considerations. Control measures shall be devised to reduce the possibility of exposure of the eye and skin to hazardous levels of laser radiation and to other hazards associated with the operation of laser devices during normal operation and maintenance.

The LSO shall have the authority to monitor and enforce the control of laser hazards and effect the knowledgeable evaluation and control of laser hazards and surveillance of the appropriate control measures.

For all uses of lasers and lasers systems, it is recommended that the minimum laser radiation level required for the application be used. Also,

---

*By permission of the Laser Institute of America, publisher of ANSI Z136.1, 12424 Research Parkway, Suite 130, Orlando, FL 32826.

it is recommended that the beam height be maintained at a level other than the normal position of the eye of a person in the standing or seated position.

Review of reported incidents have demonstrated that accidental eye and/or skin exposures to laser radiation, and accidents related to the ancillary hazards of a laser or laser system, are most often associated with personnel involved with the use of these systems under the following conditions:

1. Unanticipated eye exposure during alignment
2. Misaligned optics
3. Available eye protection not used
4. Equipment malfunction
5. Improper methods of handling high voltage
6. Intentional exposure of unprotected personnel
7. Operators unfamiliar with laser equipment
8. Lack of protection for ancillary hazards
9. Improper restoration of equipment following service

Accordingly, if alternate controls are instituted, those involved shall be provided the appropriate laser safety training (see Sections 5 and 7).

Engineering control measures (items incorporated into the laser or laser system or designed into the installation by the user) shall be given primary consideration in instituting a control measure program for limiting access to laser radiation (see 4.3).

If engineering controls are impractical or inadequate, (see 4.1.1 and 4.1.2), administrative and procedural controls (see 4.5) and personnel protective equipment (see 4.6) shall be used. The limits of any type of control measure (for example, failure modes of enclosures and eye protection, or the inability of some personnel to understand written warnings) shall be considered in developing a laser hazard control program.

4.1.1 Applicability of Control Measures. The purpose of control measures is to reduce the possibility of exposure of the eye and skin to hazardous levels of laser radiation and to associated hazards.

Whenever the application of any one or more control measures reduce the possible exposure to a level at or below the applicable MPE, the application of additional control measures for the same purpose is not required.

The control measures described in this section shall apply at all times when a laser or laser system is in its operational and maintenance modes.

If, during periods of service to a laser or laser system, the level of accessible radiation exceeds the applicable MPE, the applicable control measures shall be instituted on a temporary basis (see 4.3.12 and 4.4.8). Unless otherwise stated, the control measures shall apply to Class 3b and Class 4 lasers or laser systems. The applicability of the various control measures as related to class is shown in Table 10.

Note: The applicability of each of the following Sections with regard to the Class of the laser or laser system is designated in brackets in each Section title.

4.1.2 Substitution of Alternate Control Measures (Class 3b or Class 4). The engineering control measures recommended in 4.3 for Class 3b and Class 4 lasers or laser systems, upon review and approval by the LSO may be replaced by procedural, administrative (see 4.4) or other alternate engineering controls which provide equivalent protection. This situation could occur, for example in medical or research and development environments.

4.1.3 Lasers or Laser Systems for Medical Use (All Classes). The control measures outlined herein shall not be considered to restrict or limit in any way the use of laser radiation of any type which may be intentionally administered to an individual for diagnostic, therapeutic, or medical research purposes, by or under the direction of qualified professionals engaged in the healing arts. However, those administering and assisting in the administering of the laser radiation as well as the patient where applicable, shall be protected by the control measures as outlined herein and, as applicable, to the requirements as specified in the proposed American National Standard for Laser Safety in Health Care Facilities, ANSI Z 136.3.

4.1.4 Electrical, Beam Interaction, and other Associated Hazards. The ancillary hazards associated with the use of laser or laser systems can be significant. For example, loss of life has occurred during electrical servicing and testing of laser equipment incorporating high-voltage power supplies.

In addition, toxic fumes, particulates and intense blackbody radiation (which can present additional eye hazards) may be released during the interaction of laser energy with materials. The controls for associated hazards are included in this section, and Section 7.

4.2 Laser Safety Officer (LSO). The LSO shall have the responsibility and authority to monitor and enforce the control of laser hazards and effect the knowledgeable evaluation and control of laser hazards. This shall include, but not be limited to, such actions as establishing an NHZ, approving standard operating procedures (SOPs), avoiding unnecessary or duplicate controls, and selection of alternate controls and training. (See Sections 1, 3, 5 and Appendix D.)

4.3 Engineering Controls. The engineering control measures for a laser or laser systems are as specified in 4.3.1 to 4.3.15.

Although commercial laser products manufactured in compliance with the Federal Laser Product Performance Standard will be certified by the manufacturer and will incorporate some engineering controls, the use of the additional controls outlined in this section shall be considered in order to reduce the potential for hazard associated with some applications of lasers and laser systems.

The engineering controls (performance features) which are supplied by the manufacturer of certified products and are used in this document are described in 4.3.1, 4.3.2, 4.3.3, 4.3.4, 4.3.5.1, 4.3.7, 4.3.8, 4.3.9, and 4.3.14.

Note: Lasers and laser systems which have been altered by other than the user may necessitate recertification, reclassification and compliance reporting under the requirements of the Federal Laser Product Performance Standard.

4.3.1 Protective Housings: (All Classes). A protective housing shall be provided (except as noted in 4.3.1.1) for all classes of lasers or laser systems. (See 4.4.8) The protective housing may require interlocks (4.3.2) and labels (4.7).

4.3.1.1 Laser Use without Protective Housing: (All Classes). In some circumstances, such as research and development and during the manufacture of lasers, operation of laser or laser systems without a protective housing may become necessary. In such cases the LSO shall determine the hazard and ensure that controls are instituted appropriate to the class of maximum accessible emission to assure safe operation. These controls may include, but not be limited to:

1. access restriction (see 4.3.10)
2. eye protection (see 4.6.2)
3. area controls (see 4.3.10 and 4.3.11)

4. barriers, shrouds, beam stops etc. (see 4.3.8)

5. administrative and procedural controls (see 4.4)

6. education and training (Section 5)

4.3.2 Interlocks on Removable Protective Housings (Embedded Class 3a, Class 3b or Class 4). Protective housings which enclose embedded Class 3a, Class 3b or Class 4 lasers or laser systems shall be provided with an interlock system which is activated when the protective housing is intended to be opened during operation and maintenance. The interlock or interlock system shall be designed to prevent access to laser radiation above the applicable MPE. The interlock may, for example, be electrically or mechanically interfaced to a shutter which interrupts the beam when the protective housing is removed (see 7.4.2 for electrical design criteria).

Failsafe interlocks shall be provided for any portion of the protective housing which, by design, can be removed or displaced during normal operation and maintenance and thereby allows access to laser radiation of an embedded Class 3b or Class 4 laser. One method to fulfill the failsafe requirement is tthe use of redundant electrical series-connected interlocks.

The protective housing interlock shall not be defeated or overridden during operation.

The LSO should periodically confirm that protective housing interlocks are functional.

Adjustments or procedures during service on lasers or laser systems containing interlocks shall not cause the interlocks to be inoperative when the equipment is restored to its normal operational condition, except as noted in 4.3.1.1.

4.3.3 Service Access Panels: (Embedded Class 3b or Class 4). Portions of the protective housing which are intended to be removed only by service personnel and permit direct access to laser radiation associated with an embedded Class 3b or Class 4 laser or laser system shall either:

1. Be interlocked (fail-safe interlock not required), or

2. Require a tool for removal and shall have an appropriate warning label (see 4.7) on the panel.

If the interlock can be bypassed or defeated, a warning label with the appropriate indications shall be located on or near the interlock. The design shall not permit the service access panel to be replaced with the interlock bypassed or defeated.

4.3.4   Master Switch: (Class 3b or Class 4). A Class 3b laser or laser system should be provided with a master switch. A Class 4 laser or laser system shall be provided with a master switch. This master switch shall be operated by a key, or by a coded access (such as a computer code).

The authority for access to the master switch shall be vested in the appropriate supervisory personnel. The master switch shall be disabled when the laser or laser system is not intended to be used.

4.3.5   Viewing Portals and Collecting Optics

4.3.5.1   Viewing Portals and Display Screens: (Class 2, Class 3a, Class 3b or Class 4). All viewing portals and/or display screens included as an integral part of a Class 2, Class 3a, Class 3b or Class 4 laser or laser system shall incorporate a suitable means (such as interlocks, filters, attenuators) to maintain the laser radiation at the viewing position at or below the applicable MPE for all conditions of operation and maintenance.

4.3.5.2   Collecting Optics (Class 2, Class 3a, Class 3b or Class 4). All collecting optics (such as lenses, telescopes, microscopes, endoscopes, etc.), intended for viewing use with a laser or laser system shall incorporate suitable means (such as interlocks, filters, attenuators) to maintain the laser radiation transmitted through the collecting optics to levels at or below the appropriate MPE, under all conditions of operation and maintenance. Note: Normal or prescription eyewear are not considered collecting optics.

4.3.6   Beam Paths: (Class 3b or Class 4). Control of the laser beam path shall be accomplished as described in the following sections.

4.3.6.1   Totally Unenclosed Beam Path: (Class 3b or Class 4). In applications of Class 3b or Class 4 lasers or laser systems where the entire beam path is unenclosed, a laser hazard analysis shall be effected by the LSO to establish the NHZ if not furnished by the manufacturer or available as part of the classification (see 3.4).

The analysis will define the area where laser radiation is accessible at levels above the appropriate MPE and will define the zone requiring control measures. A laser controlled area shall be established in this zone (see 4.3.10 and 4.3.11) and appropriate control measures shall be implemented based upon the classification associated with the maximum level of accessible laser radiation.

4.3.6.2   Limited Open Beam Path: (Class 3b or Class 4). In applications of Class 3b or Class 4 lasers or laser systems where the beam path is confined by design to significantly limit the degree of accessibility of the open beam, a hazard analysis shall be effected by the LSO to establish the NHZ (see 3.4). The LSO shall establish controls appropriate to the magnitude and extent of the accessible radiation.

Frequently the hazard analysis will define an extremely limited NHZ and procedural controls can provide adequate protection.

4.3.6.3 Enclosed Beam Path: (All Classes). In applications of laser or laser systems where the entire beam path is enclosed, and the enclosure fulfills all requirements of a protective housing (i.e., limits the exposure to the MPE; see 4.3.1 and 4.3.2), the requirements of Class 1 are fulfilled and no further controls are required.

When the protective housing requirements are temporarily relaxed, such as during service, the LSO shall establish the appropriate controls. These may include a temporary area control (see 4.3.12) and administrative and procedural controls (see 4.4).

4.3.7 Remote Interlock Connector: (Class 3b or Class 4). A Class 3b laser or laser system should be provided with a remote interlock connector. A Class 4 laser or laser system shall be provided with a remote interlock connector.

The interlock connector facilitates electrical connections to an emergency master disconnector interlock, or to a room, entryway, or area interlock, as may be required for a Class 4 controlled area (see 4.3.10.2).

When the terminals of the connector are open circuited, the accessible radiation shall not exceed the appropriate MPE levels.

4.3.8 Beam Stop or Attenuators (Class 3a, Class 3b or Class 4). A Class 3a or a Class 3b laser or laser system should be provided with a permanently attached beam stop or attenuator.

A Class 4 laser or laser system shall be provided with a permanently attached beam stop or attenuator.

The beam stop or attenuator shall be capable of preventing access to laser radiation in excess of the appropriate MPE level when the laser or laser system output is not required.

There are many instances, such as during service, when a temporary beam attenuator placed over the beam aperture can reduce the level of accessible laser radiation to levels below the applicable MPE level. Such a practice can replace the requirements for laser eye protection for some critical alignment procedures.

4.3.9 Laser Activation Warning Systems: (Class 3b or Class 4). An alarm (for example, an audible sound such as a bell or chime), a warning light (visible through protective eyewear), or a verbal "countdown" command should be used with Class 3b, and shall be used with Class 4 lasers or laser systems during activation or startup.

Distinctive and clearly identifiable sounds that arise from auxiliary equipment (such as a vacuum pump or fan) and that are uniquely associated with the emission of laser radiation are acceptable as an audible warning.

4.3.9.1 Emission Delay (Class 4). For Class 4 lasers or laser systems, the warning system shall be activated a sufficient time prior to emission of laser radiation to allow appropriate action to be taken to avoid exposure to the laser radiation.

4.3.10 Indoor Laser Controlled Area (Class 3b or Class 4). A laser hazard analysis, including determination of the NHZ, shall be effected by the LSO. If the analysis determines that the classification associated with the maximum level of accessible radiation is Class 3b or Class 4, a laser controlled area shall be established and adequate control measures instituted (see 4.3.10.1 or 4.3.10.2).

Note: The requirements for nonenclosed lasers or laser systems involving the general public are detailed in 4.5.1.

4.3.10.1 Class 3b Laser Controlled Area (Class 3b). The Class 3b laser controlled area shall:

1. Be posted with the appropriate warning sign(s) (see 4.7), except as detailed in 4.5.1.10
2. Be operated by qualified and authorized personnel
3. Be operated in a manner such that the path is limited when the NHZ of the laser beam must exit an indoor controlled area, particularly to the outdoors under adverse atmospheric conditions, i.e., rain, fog, snow, etc. See 4.3.11 or 7.6

In addition to the above, a Class 3b controlled area should:

1. Be under the direct supervision of an individual knowledgeable in laser safety
2. Be so located that access to the area by spectators is limited and requires approval, except as detailed in 4.5
3. Have any potentially hazardous beam terminated in a beam stop of an appropriate material

4. Have only diffuse reflective materials in or near the beam path, where feasible

5. Have personnel within the controlled area provided with the appropriate eye protection if there is any possibility of viewing the direct or reflected beams

6. Have the laser secured such that the beam path is above or below eye level of a person in any standing or seated position, except as required for medical use

7. Have all windows, doorways, open portals, etc. from an indoor facility be either covered or restricted in such a manner as to reduce the transmitted laser radiation to levels at or below the appropriate ocular MPE

8. Require storage or disabling (for example, removal of the key) of the laser or laser system when not in use to prevent unauthorized use

4.3.10.2 Class 4 Laser Controlled Area (Class 4). All personnel who regularly require entry into a Class 4 laser controlled area shall be adequately trained, provided with appropriate protective equipment, and follow all applicable administrative and procedural controls.

All Class 4 area/entryway safety controls shall be designed to allow both rapid egress by laser personnel at all times and admittance to the laser controlled area under emergency conditions.

For emergency conditions there shall be a clearly marked "Panic Button" (remote controlled connector or equivalent device) available for deactivating the laser or reducing the output to the appropriate MPE levels.

The Class 4 laser controlled area shall be designed to fulfill all items of 4.3.10.1, and in addition shall incorporate one of the following alternatives.

1. Non-Defeatable (non-override) Area Entryway Safety Controls.

2. Non-defeatable safety latches, entryway or area interlocks (e.g., electrical switches, pressure sensitive floor mats, infrared or sonic detectors) shall be used to deactivate the laser or reduce the output to the appropriate MPE levels in the event of unexpected entry into the laser controlled area. (See Figure 7.2a).

3. Defeatable Area/Entryway Safety Controls. Defeatable safety latches, entryway or area interlocks shall be used if non-defeatable area/entryway safety controls limit the intended use of the laser

or laser system. For example, during normal usage requiring operation without interruption (e.g., long term testing, medical procedures, surgery), if it is clearly evident that there is no optical radiation hazard at the point of entry, override of the safety controls shall be permitted to allow access to authorized personnel provided that they have been adequately trained and provided with adequate personal protective equipment. (See Figure 7.2a.)

4. Procedural Area/Entryway Safety Controls. Where safety latches or interlocks are not feasible or are inappropriate, for example during medical procedures, surgery, etc., the following shall apply (see Fig. 7.2b):

   (a) All authorized personnel shall be adequately trained and adequate personal protective equipment shall be provided upon entry

   (b) A door, blocking barrier, screen, curtains, etc. shall be used to block, screen or attenuate the laser radiation at the entryway. The level of laser radiation at the exterior of these devices shall not exceed the applicable MPE, nor shall personnel experience any exposure above the MPE immediately upon entry

   (c) From the entryway there shall be a visible or audible signal indicating that the laser is energized and operating at Class 4 levels. A lighted laser warning sign or flashing light are two of the appropriate methods to accomplish this requirement

4.3.11 Outdoor Control Measures (Class 3b or Class 4).* A Class 3b or Class 4 laser or laser system used outdoors shall meet the following requirements:

1. The LSO shall effect an analysis to establish the NHZ if not available as part of the classification documentation nor furnished by the manufacturer

2. Unprotected, untrained and unauthorized personnel shall be excluded from the beam path at all points where the appropriate MPE is exceeded

3. Appropriate combinations of physical barriers, screening, protective eye and body wear or appropriate administrative controls shall be used if personnel are permitted within the NHZ

---

*The laser as an outdoor art form is discussed in D. C. Winburn, *What Every Engineer Should Know About Lasers*, Marcel Dekker, New York, 1987.

4. Directing the laser beam toward automobiles, aircraft or other manned vehicles shall be prohibited within the NHZ

5. The laser beam path shall not be maintained at or near personnel eye level without specific authorization of the LSO

6. The beam path shall be terminated where possible

7. When the laser is not being used, it shall be disabled in a manner that prevents unauthorized use

8. Only qualified and authorized personnel shall operate the laser or laser system

9. The operation of Class 4 lasers or laser systems during rain or snowfall or in fog or dusty atmosphere may produce hazardous reflections near the beam. In such applications, the LSO shall effect an evaluation of the need for, and specify the use of, appropriate personal protective equipment

4.3.13 Remote Firing and Monitoring (Class 4). Whenever appropriate and possible, Class 4 lasers or laser systems should be monitored and fired from remote positions. The remote console should also include a laser activation warning system.

4.3.12 Temporary Laser Controlled Area (Embedded Class 3b or Class 4). In those conditions where removal of panels or protective housings, over-riding of protective housing interlocks, or entry into the NHZ becomes necessary (such as for service), and the accessible laser radiation exceeds the applicable MPE, a temporary laser controlled area shall be devised for an embedded Class 3b or Class 4 laser or laser system.

Such an area, which by its nature will not have the built-in protective features as defined for a laser controlled area, shall provide all safety requirements for all personnel, both within and without.

A notice sign (see 4.7.4.3) shall be posted outside the temporary laser controlled area to warn of the potential hazard.

4.3.14 Equipment Labels: (All Classes Except Class 1). All lasers or laser systems (except Class 1) shall have appropriate warning labels (see 4.7) with the appropriate cautionary statement. The label shall be affixed to a conspicuous place on the laser housing or control panel. Such labels should be placed on both the housing and the control panel if these are separated by more than 2 meters.

Note: A Class 2a laser requires a label, but no symbol. Additional wording for non-laser hazards may be required (see 7.4.6).

4.3.15 Area Posting Signs (Class 2, Class 3a, Class 3b or Class 4). An area which contains a Class 2 or Class 3a laser or laser system should be posted with the appropriate sign as described in 4.7, except as noted in 4.5.1.10.

An area which contains a Class 3b or Class 4 laser or laser system shall be posted with the appropriate sign as described in 4.7.

A notice sign, as described in 4.7, shall be posted outside a temporary laser control area.

4.4 Administrative and Procedural Controls. Administrative and procedural controls are methods or instructions which specify rules, or work practices, or both, which implement or supplement engineering controls and which may specify the use of personal protective equipment. Unless otherwise specified, administrative and procedural controls shall apply only to Class 3b and Class 4 lasers or laser systems.

4.4.1 Standard Operating Procedures: (Class 3b or Class 4). The LSO should require approved written standard operating, maintenance and service procedures (SOP's) for Class 3b lasers or laser systems. The LSO shall require and approve written SOP's for Class 4 lasers or laser systems. These written SOP's shall be maintained with the laser equipment for reference by the operator, and maintenance or service personnel.

4.4.2 Output Emission Limitations: (Class 3a, Class 3b or Class 4). If, in the opinion of the LSO, excessive power or radiant energy is accessible during operation and maintenance, the LSO shall take such action as required to reduce the levels of accessible power or radiant energy to that which is commensurate with the required application.

4.4.3 Education and Training: (Class 2, Class 3a, Class 3b or Class 4). Education and training shall be provided for operators, maintenance or service personnel for Class 3a, Class 3b or Class 4 lasers or laser systems. Education and training should be provided for operators, maintenance or service personnel for Class 2 lasers or laser systems. The level of training shall be commensurate with the level of potential hazard (see Section 5 and Appendix D).

4.4.4 Authorized Personnel (Class 3b or Class 4). Class 3b or Class 4 lasers or laser systems shall be operated, maintained or serviced only by authorized personnel. Lasers or laser systems with embedded Class 3b or Class 4 lasers shall be maintained or serviced only by authorized personnel if such procedures would permit access to levels which exceed the appropriate MPE.

4.4.5 Alignment Procedures (Class 2, Class 3a, Class 3b or Class 4).
Alignment of laser optical systems (mirrors, lenses, beam deflectors, etc.)
shall be performed in such a manner that the primary beam, or a specular or diffuse reflection of a beam, does not expose the eye to a level
above the applicable MPE.

Written procedures outlining alignment methods should be approved
for Class 3b and shall be approved for Class 4 lasers or laser systems. The
use of low power (Class 1 or Class 2) visible lasers for path simulation
of higher power lasers is recommended for alignment of higher power
Class 3b or Class 4 visible or invisible lasers and laser systems.

Experience has shown that a significant ocular hazard may exist during alignment procedures.

4.4.6 Eye Protection (Class 3b or Class 4). Eye protection devices
which are specifically designed for protection against radiation from
Class 3b lasers or laser systems should be administratively required and
their use enforced when engineering or other procedural and administrative controls are inadequate to eliminate potential exposure in excess of
the applicable MPE (see 4.6).

Eye protection devices which are specifically designed for protection
against radiation from Class 4 lasers of laser systems shall be administratively required and their use enforced when engineering or other procedural and administrative controls are inadequate to eliminate potential
exposure in excess of the applicable MPE (see 4.6).

When long-term exposure to visible lasers is not intended, the applicable MPE used to establish the optical density requirement for eye protection may be based on a 0.25 second exposure.

When viewing an extended source or the diffuse reflection of the beam
from a Class 3b or Class 4 laser or laser system where intermediate viewing time is intended, e.g., optical alignment procedures, the applicable
MPE should be based upon the maximum viewing time anticipated during any given 8 hour period. For example, a maximum viewing time of
600 seconds is applicable for most alignment procedures when viewing
a diffusely reflecting target.

If the extended source criteria (8.3) is not applicable due to a small
beam size, then the exposure at the cornea from a diffuse reflector may
be estimated using the inverse square law relationship and the intrabeam
MPE criteria shall apply.

4.4.7 Spectators (Class 3b or Class 4). Spectators should not be permitted within a laser controlled area (see 4.3.10 and 4.3.11) which contains a Class 3b laser or laser system and spectators shall not be permitted

within a laser controlled area which contains a Class 4 laser or laser system unless:

1. appropriate supervisory approval has been obtained
2. the degree of hazard and avoidance procedure has been explained
3. appropriate protective measures are taken

Laser demonstrations involving the general public shall be governed by the requirements of 4.5.

4.4.8 Service Personnel (All Classes). Personnel who require access to Class 3b or Class 4 lasers or laser systems contained within a protective housing or protected area enclosure for the purpose of service shall comply with the appropriate control measures of the enclosed or embedded laser or laser system. The service personnel shall have the education and training commensurate with the class of the laser or laser system contained within the protective housing.

4.5 Special Considerations

4.5.1 Laser Demonstrations Involving the General Public. The following special control measures shall be employed for those situations where lasers or laser systems are used for demonstration, artistic display, entertainment or other related uses where the intended viewing group is the general public.

Such demonstrations can be, but are not limited to, trade show demonstrations, artistic light performances, planetarium laser shows, discotheque lighting, stage lighting effects and similar special lighting effects that use lasers or laser systems emitting in the visible wavelength range (0.4 to 0.7 μm).

For the purposes of this section, the applicable MPE may be determined by using the classification duration defined as the total combined operational time of the laser during the performance or demonstration within any single period of 3×10^4 seconds.

4.5.1.1 Operational Requirements—General. Only Class 1 laser or laser systems shall be used for general public demonstration, display, or entertainment in unsupervised areas without additional requirements.

If unsupervised, the use of Class 2 and Class 3a lasers or laser systems shall be limited to installations which prevent access to the direct or specularly reflected beams or where the accessible radiation is maintained at the distance requirements specified in 4.5.1.6.

The use of Class 3b or Class 4 lasers or laser systems shall be permitted only under the following conditions:

1. when the laser operation is under the control of an experienced, trained operator
2. when the laser is operated in a supervised laser installation as specified in 4.5.1.7
3. when the laser is operated in an unsupervised laser installation provided: (a) the LSO has assured that appropriate controls are incorporated which provide for an equivalent degree of protection to the general public; (b) a designated person, present at all times at the show or display, is responsible for the immediate termination of the laser equipment in the event of equipment malfunction, audience unruliness, or other unsafe conditions.

4.5.1.2 Invisible Laser Emission Limitations. The general public shall not be exposed nor have access to laser radiation emission at wavelengths outside the visible range (0.4 to 0.7 μm) at levels exceeding the applicable MPE levels under any possible conditions of operation.

4.5.1.3 Optical Radiation Limits for the General Public. Laser radiation in the location where the general public is normally allowed shall not exceed the applicable MPE levels during operation. Laser radiation to be considered shall include reflections from all possible surfaces and scattering materials.

4.5.1.4 Operator and Performers. All operators, performers, and employees shall be able to perform their required functions without the need for exposure to laser radiation levels in excess of the applicable MPE level.

4.5.1.5 Scanning Devices. Scanning devices, including rotating mirrored balls, shall incorporate a means to prevent laser emission if scan failure or other failure resulting in a change in either scan velocity or amplitude would result in failure to fulfill the criteria given in 4.5.1.3 and 4.5.1.4.

4.5.1.6 Unsupervised Installations—Distance Requirements. If a laser demonstration using a Class 2, Class 3a, Class 3b or Class 4 laser does not operate at all times under the direct supervision or control of an experienced, trained operator, the laser radiation levels to which access can be gained shall be limited by barriers, windows, or other means so as not to exceed the limits of the applicable MPE. This limitation applies at any point less than 6 m above any surface upon which a person in the general public is permitted to stand and to any point less than 2.5 m in lateral separation from any position where a person in the general public is permitted to be present during the performance or display. See Figures 2c, 2d and 2e.

4.5.1.7 Supervised Laser Installations. Laser demonstrations which do not meet the criteria stated in 4.5.1.6 shall be operated at all times under the direct supervision or control of an experienced, trained operator who shall maintain constant surveillance of the laser display and terminate the laser emission in the event of equipment malfunction, audience unruliness or other unsafe conditions.

The operator shall have visual access to the entire area of concern. If obstacles or size preclude visual access by the operator, the multiple observers shall be used, with a communication link to the operator. In such supervised installations accessible laser radiation shall be limited by barriers, windows, or other means so as to not exceed the applicable MPE (see 4.5.1) at any point unless the following requirements are met:

1. The accessible laser radiation is maintained a minimum distance of 3.0 m above any surface upon which the general public would be able to stand during a performance
2. The accessible laser radiation is maintained a minimum distance of 2.5 m in lateral separation from any position where the general public is permitted to be present during the performance or demonstration.

As a general practice, the largest practical vertical and lateral separation distances should be employed wherever possible. However, if specific physical limitations of the space in which the lasers are to be used preclude compliance with the required minimum separations specified in (1) and (2), shorter separation distances may be utilized, provided that the LSO assures that alternative measures are established which assure an equivalent degree of protection to the general public.

4.5.1.8 Beam Termination Requirements. All laser demonstration systems shall be provided with a readily accessible means to effect immediate termination of the laser radiation. If the demonstration does not require continuous supervision or operator control during its operation, there must be a designated person at all times at the show or display who is responsible for the immediate termination of the laser radiation in the event of equipment malfunction, audience unruliness or other unsafe conditions.

4.5.1.9 Maximum Power Limitations. The maximum output power of the laser shall be limited to the level required to produce the desired and intended effect within the limitations outlined in 4.5.1.1 through 4.5.1.8.

4.5.1.10 Posting of Warning Signs and Logos. If the laser installation fulfills all requirements as detailed in 4.5.1.1 through 4.5.1.9—or as alternatively provided by the LSO (see, for example, 4.5.1.7)—then posting of area warnings signs shall not be required.

4.5.1.11 Federal, State, or Local Requirements. The laser operator and/or LSO responsible for producing the laser demonstration should determine that any applicable federal, state, or local requirements are satisfied.

4.5.2 Laser Optical Fiber Transmission System. Laser transmission systems which employ optical cables shall be considered enclosed systems with the optical cable forming part of the enclosure. If disconnection of a connector results in accessible radiation being reduced to below the applicable MPE by engineering controls, connection or disconnection may take place in an uncontrolled area and no other control measures are required. When the system provides access to laser radiation above the applicable MPE via a connector, the conditions given in 4.5.2.1 or 4.5.2.2, or for optical communication systems as specified in the proposed American National Standard for the Safe Use of Optical Fiber Communication Systems Utilizing Laser Diode and LED Sources, ANSI Z136.2, shall apply.

4.5.2.1 Connection or disconnection during operation shall take place in an appropriate laser controlled area, in accordance with the requirements of 4.3.12, if the MPE is exceeded.

4.5.2.2 Connection or disconnection during maintenance, modification, or service shall take place in a temporary laser controlled area, in accordance with 4.3.12 if the MPE is exceeded. When the connection or disconnection is made by means of a connector, other than one within a secured enclosure, such a connector shall be disconnectable only by the use of a tool.

When the connection or disconnection is made within a secured enclosure, no tool for connector disconnection shall be required, but a warning sign appropriate to the class of laser or laser system shall be visible when the enclosure is open.

4.6 Personal Protective Equipment

4.6.1 General. Enclosure of the laser equipment or beam path is the preferred method of control, since the enclosure will isolate or minimize the hazard.

When other control measures do not provide adequate means to prevent access to direct or reflected beams at levels above the MPE, it may be necessary to use personal protective equipment. It should be noted that personal protective equipment may have serious limitations when used as the only control measure with higher-power Class 4 lasers or laser systems; the protective equipment may not adequately reduce or eliminate the hazard.

4.6.2 Protective Eyewear (Class 3b or Class 4)

4.6.2.1 Class 3b Laser. Protective eyewear should be worn whenever operational conditions may result in a potential eye hazard.

4.6.2.2 Class 4 Laser. Protective eyewear shall be worn whenever operational conditions may result in a potential eye hazard.

4.6.2.3 Eyewear for Protection Against Other Agents. Physical and chemical hazards to the eye can be reduced by the use of face shields, goggles, and similar protective devices. Consult American National Standard Practice for Occupational and Educational Eye and Face Protection, ANSI Z87.1-1979.

4.6.2.4 Factors in Determining Appropriate Eyewear. The following factors shall be considered in determining the appropriate protective eyewear to be used:

1. Wavelength of laser output
2. Potential for multiwavelength operation
3. Radiant exposure or irradiance
4. Maximum permissible exposure (MPE)
5. Optical density of eyewear at laser output wavelength
6. Visible light transmission requirement

7. Peripheral vision requirement

8. Radiant exposure or irradiance and the corresponding time factors at which laser safety eyewear damage (penetration) occurs, including transient bleaching

9. Need for prescription glasses

10. Comfort and fit

11. Degradation of absorbing media, such as photobleaching

12. Strength of materials (resistance to shock)

13. Capability of the front surface to produce a specular reflection

4.6.2.5 Optical Density

4.6.2.5.1 Specification of Optical Density, D_λ. The optical density (attenuation), D_λ, of laser protective eyewear at a specific wavelength shall be specified. Many lasers radiate at more than one wavelength; thus eyewear designed to have an adequate D_λ for a particular wavelength could have a completely inadequate D_λ at another wavelength radiated by the same laser. This problem may become particularly serious with lasers that are tunable over broad wavelength bands. In such cases, alternative methods of eye protection, such as indirect viewing, may be more appropriate (e.g., image converters, closed circuit TV).

If the potential eye exposure is given by H_p, the optical density, D_λ, required of protective eyewear to reduce this exposure to the MPE is given by

$$D_\lambda = \log_{10} \frac{H_p}{\text{MPE}} = -\log_{10} T_\lambda$$

where H_p is expressed in the same units as the appropriate MPE and T_λ is the transmittance of the filter at the specific wavelength.

Note: When the laser beam is smaller than the limiting aperture (L_A), the value of H_p is determined by averaging the beam energy over the limiting aperture (7 mm for the 0.4 to 1.4 μm region). This is required because the MPE value has been established (normalized) relative to the limiting aperture area. Use of the beam diameter (a) to evaluate H_p for cases where $a < L_A$ will result in excessively large D_λ specifications.

If the beam size is larger than the size of the protective eyewear, it should be noted that optical densities greater than 3 or 4 (depending upon exposure time) could reduce eye exposure below the MPE, but leave the unprotected skin surrounding the eyewear exposed to values in excess of the skin MPE.

Attenuation through the protective material shall be determined for all anticipated viewing angles and wavelengths.

4.6.2.5.2 Optical Density for Non-Laser Emissions. Eye protection shall be provided for the ultraviolet and blue-green spectral region (0.2 to 0.55 μm) for laser welding processes. On the basis of currently available data, a minimum optical density, D_λ, of 2.0 (neutral density) or welding shade 7 (see American National Standard Practice for Occupational and Educational Eye and Face Protection, ANSI Z87.1-1979 or the latest revision thereof) is recommended for emission in the ultraviolet and blue-green spectral regions. The optical density values given above would apply for the laser induced plasma, such as that associated with typical laser welding systems. Greater optical densities are required for higher power laser welding systems.

The plasma emission optical density specified does not replace the optical density requirements for laser emission as specified in 4.6.2.5.1.

4.6.2.6 Visible Transmission. Adequate optical density, D_λ, at the laser wavelength of interest shall be weighted with the need for adequate visible transmission.

4.6.2.7 Identification of Eyewear. All laser protective eyewear shall be clearly labeled with the optical density and wavelength for which protection is afforded (see 4.6.2.5). Color coding or other distinctive identification of laser protective eyewear is recommended in multi-laser environments.

4.6.2.8 Comfort and Fit. Protective eyewear should be comfortable and prevent exposure to hazardous peripheral radiation.

4.6.2.9 Inspection. Periodic inspection shall be made of protective eyewear to ensure the maintenance of satisfactory condition. This shall include:

1. Inspection of the attenuation material for pitting, crazing, cracking, discoloration etc.
2. Inspection of the frame for mechanical integrity
3. Inspection for light leaks that would permit hazardous intrabeam viewing

Eyewear in suspicious condition should be tested for acceptability or discarded.

4.6.2.10 Purchasing Information for Protective Eyewear. Purchasers of laser safety protective eyewear should ensure that the following information accompanies each item:

1. Wavelength(s) and corresponding optical density for which protection is afforded
2. Pertinent data for laser safety purposes

4.6.3 Protective Clothing. Where personnel may be exposed to levels of radiation that clearly exceed the MPE for the skin, particularly in the ultraviolet, the LSO shall recommend or approve the use of protective clothing. Where personnel may be subject to chronic skin exposure from scattered ultraviolet radiation, as may occur during excimer laser processing, skin protection should be provided even at levels below the MPE for the skin. Consideration should also be given to the use of fire resistant material when using Class 4 lasers.

4.6.4 Other Personnel Protective Equipment. Respirators and hearing protection may be required whenever engineering controls cannot provide protection from a harmful ancillary environment (see 7.1).

4.7 Warning Signs and Labels

4.7.1 Design of Signs. Sign dimensions, letter size and color, etc., shall be in accordance with American National Standard Specification for Accident Prevention Signs, ANSI Z35.1-1972 (or the latest revision thereof).

4.7.2 Symbols. The laser hazard symbol shall be a sunburst pattern consisting of two sets of radial spokes of different lengths and one long spoke, radiating from a common center.

The color, dimensions, and location of the symbol within the sign shall be as specified in the American National Standard Specification for Accident Prevention Signs, ANSI Z35.1-1972, (or the latest revision thereof).

Note: Classification labeling in accordance with the Federal Laser Product Performance Standard may be used to satisfy the labeling requirements in this section.

4.7.3 Signal Words

4.7.3.1 The signal word "Caution" shall be used with all signs and labels associated with Class 2 lasers and laser systems and all Class 3a lasers and laser systems that do not exceed the appropriate MPE for irradiance. The signal word "Danger" shall be used with all signs and labels associated with all other Class 3a, and all Class 3b and Class 4 lasers and laser systems.

4.7.3.2 A Class 2a laser or laser system shall have a label affixed which instructs: "Avoid Long-Term Viewing of Direct Laser Radiation". This label need not bear the warning symbol (see 4.7.2) or signal words but must be clearly visible during operation and bear the designation "Class 2a Laser".

4.7.3.3 The word "Radiation" on signs and labels may be replaced by the word "Light" for lasers operating in the visible range at wavelengths greater than 400 nm and equal to or less than 700 nm.

For lasers operating outside of this visible range the word "invisible" shall be placed prior to the word "radiation."

4.7.4 Inclusion of Pertinent Information. Signs and labels shall conform to the following specifications.

4.7.4.1 The appropriate signal word (Caution or Danger) shall be located in the upper panel.

4.7.4.2 Adequate space shall be left on all signs and labels to allow the inclusion of pertinent information.

Such information may be included during the printing of the sign or label or may be handwritten in a legible manner, and shall include the following:

1. At position 1 above the tail of the sunburst, special precautionary instructions or protective actions required by the reader such as:
   a. For Class 2 and Class 3a lasers and laser systems where the accessible irradiance does not exceed the appropriate MPE based upon a 0.25 second exposure (see 3.2.3.3.3) "Laser Radiation — Do Not Stare into Beam or View with Optical Instruments"
   b. For all other Class 3a lasers and laser systems, "Laser Radiation — Avoid Direct Eye Exposure"

    c. For all Class 3b lasers and laser systems, "Laser Radiation — Avoid Direct Exposure to Beam"

    d. For Class 4 lasers and laser systems, "Laser Radiation — Avoid Eye or Skin Exposure to Direct or Scattered Radiation"

2. At position 1 above the tail of the sunburst, special precautionary instructions or protective action such as: Invisible Laser Radiation; Knock Before Entering; Do Not Enter When Light is On; Restricted Area; etc.

3. At position 2 below the tail of the sunburst, type of laser (Ruby, Helium-Neon, etc.), or the emitted wavelength, pulse duration (if appropriate), and maximum output

4. At position 3, the class of the laser or laser system

4.7.4.3 Temporary Laser Controlled Area Signs: (Class 3b or Class 4). A notice sign (see Figure 1c) shall be posted outside a temporary laser controlled area (see 4.3.12), for example, during periods of service. When a temporary laser controlled area is created, the area outside the temporary area remains Class 1, while the area within is either Class 3b or Class 4 and the appropriate danger warning is required. (See Figure 1b.)

4.7.4.4 Display Signs and Labels. All signs and labels shall be conspicuously displayed in locations where they best will serve to warn onlookers.

4.7.4.5 Existing Signs and Labels. Signs and labels prepared in accordance with previous revisions of this standard are considered to fulfill the requirement of the standard.

4.8 Service and Repair of Laser Systems. Following any service or repair which may affect the output power or operating characteristics of a laser or laser system so as to make it potentially more hazardous, the LSO shall ascertain whether any changes are required in control measures.

4.9 Modification of Laser Systems. Whenever deliberate modifications are made which could change a laser or laser system's class and affect its output power or operating characteristics so as to make it potentially more hazardous, the LSO shall ascertain whether any changes and control measures are required.

8

Selecting Laser-Protective Eyewear

The eye is susceptible to damage from low intensity levels of laser light. In this chapter we will present considerable information to ensure that personnel working in a laser environment understand the criteria in selecting eye protective devices. In addition, several myths will be attacked with factual argument so that the topic of management of eye protection can be approached with confidence in providing workers with simple, effective, lightweight, and comfortable eyewear that will also be cost effective. The importance of this chapter cannot be overemphasized.

First, we will discuss some of the circumstances surrounding some of the recorded eye damage accidents involving lasers. In all reported cases, except two, of permanent retinal damage read by the author in current literature, permanent damage resulted from the injured person *not wearing any protective eyewear!* The exceptions involved (1) a laser user who "peeked" under the frame, and (2) a user who relied on excessive advertized optical density for protection (see page 115). The following case histories are presented to substantiate the simple fact that eyewear was not worn, and to expose the fact that extremely low energy densities of pulsed lasers can cause retinal damage. (It should be noted here that corneal damage recorded in literature is rare, mainly because, in the author's opinion, any wavelength in the infrared (IR) or ultraviolet (UV) ranges that could cause corneal damage is absorbed by plastic and glass materials worn routinely in a laboratory or industrial environment, and laser intensities required to cause corneal damage are quite high, comparatively.)

CASE HISTORY NO. 1

This is an accident victim's viewpoint of his experience.*

The necessity for safety precautions with high-power lasers was forcibly brought home to me last January when I was partially blinded by a reflection from a relatively weak neodymium-yag laserbeam. Retinal damage resulted from a 6-millijoule, 10-nanosecond pulse of invisible 1064-nanometer radiation. *I was not wearing protective goggles* at the time, although they were available in the laboratory. As any experienced laser researcher knows, goggles not only cause tunnel vision and become fogged, they become very uncomfortable after several hours in the laboratory.

When the beam struck my eye I heard a distinct popping sound, caused by a laser-induced explosion at the back of my eyeball. My vision was obscured almost immediately by streams of blood floating in the vitreous humor, and by what appeared to be particulate matter suspended in the vitreous humor. It was like viewing the world through a round fishbowl full of glycerol into which a quart of blood and a handful of black pepper have been partially mixed. There was local pain within a few minutes of the accident, but it did not become excruciating. The most immediate response after such an accident is horror. As a Vietnam War Veteran, I have seen several terrible scenes of human carnage, but none affected me more than viewing the world through my bloodfilled eyeball. In the aftermath of the accident I went into shock, as is typical in personal injury accidents.

As it turns out, my injury was severe but not nearly as bad as it might have been. I was not looking directly at the prism from which the beam had reflected, so the retinal damage is not in the fovea. The beam struck my retina between the fovea and the optic nerve, missing the optic nerve by about three millimeters. Had the focused beam struck the fovea, I would have sustained a blind spot in the center of my field of vision.

---

*Comment, *Laser Focus*, August 1977.

Figure 8.1 This is the same diagram as shown in Chapter 3, Figure 3.4; it is repeated here to emphasize location of damage caused by laser radiation and to introduce the first of 10 case histories of recorded laser eye damage.

Had it struck the optic nerve, I probably would have lost the sight of that eye.

The beam did strike so close to the optic nerve, however, that it severed nerve-fiber bundles radiating from the optic nerve. This has resulted in a cresent-shaped blind spot many times the size of the lesion. The diagram [Figure 8.1] is a Goldman-Fields scan of the damaged eye, indicating the sightless portions of my field of view four months after the accident. The small blind spot at the top exists for no discernible reason;

the lateral blind spot is the optic nerve blind spot. The effect of the large blind area is much like having a finger placed over one's field of vision. Also, I still have numerous floating objects in the field of view of my damaged eye, although the blood streamers have disappeared. These "floaters" are more a daily hindrance than the blind areas, because the brain tries to integrate out the blind area when the undamaged eye is open. There is also recurrent pain in the eye, especially when I have been reading too long or when I get tired.

The moral of all this is to be careful and to wear protective goggles when using highpower lasers. The temporary discomfort is far less than the permanent discomfort of eye damage. The type of reflected beam which injured me also is produced by the polarizers used in q switches, by intracavity diffraction gratings, and by all beamsplitters or polarizers used in optical chains.

The victim of Case History No. 1 explained that protective goggles were not being worn at the time, although they were available in the laboratory. He also stated, "As any experienced laser researcher knows, goggles not only cause tunnel vision and become fogged, they become very uncomfortable after several hours in the laboratory." This statement substantiates the author's contention that management of eye protection must include knowledge of all available types of eyewear and not accept the myth that heavy, uncomfortable goggles must be worn to ensure safety. This chapter will provide knowledge of available types and styles of devices, particularly comparing spectacles versus goggles.

CASE HISTORY NO. 2

Another viewpoint from a victim describing the circumstances of an accident follows:*

As I read my November issue of *Laser Focus* I took note of the eye injury report, curious about the particulars of this novel accident. Even though I have been working with lasers for five years in the presence of many of the same hazards pointed to in the article, I didn't think while reading it, "This could happen to me!" But it did.

Laser Focus, 1982

On January 22, 1982, I spent several hours aligning a low-power, frequency-doubled Nd:YAG beam through a dye laser set-up. In order to see the 532-nm pump beam propagation *I was not wearing goggles*. I had also removed a beam block intended to absorb a Brewster's angle reflection, to observe end pumping of an emplifier cell. The green power was increased to determine the extent of dye lasing without replacing the beam block. I did not put on goggles. While placing a power meter at the dye laser output I leaned over the uncovered amplifier and caught a reflection in my right eye. Because I was in continuous motion looking at the meter and not the beam, I doubt that more than one 10 to 15 nsec pulse of ~20 microjoules was focused onto the fovea. While I do remember seeing a green flash there was no pain. I was not immediately aware of any significant eye damage. It wasn't until I shut the lasers off and returned to my desk to record the day's activity that I realized I had a blind spot comparable to a comera flash, but only in my right eye. It was almost 5:00 p.m. on a Friday, and I didn't report the incident because I couldn't believe that any serious damage was done.

By Saturday afternoon I knew I had a problem. Monday the 25th I notified our safety division and started my visits to an ophthalmologist. The initial examination supported the probability of permanent damage although hemorrhaging in the affected area obstructed detail. By the end of the first week, peripheral vision around the spot was improving (due to decreased swelling), and the actual contact point was observed to be on the right side of the macula (that corresponds to a blind spot slightly off center left). I was encouraged and felt fortunate considering the negative potential of this careless mistake. But by week two peripheral vision had declined. Distortion (curving) of resolution around the spot became more noticeable due to additional blood pooling under the retina. If this hemorrhaging were to persist, laser cauterization would be necessary. But for now, "treatment" consists of waiting, observing, and photographing.

Although recovery has not been straightforward, and my vision may get worse before it gets better, I still feel lucky in that one eye totally escaped injury. So while reading was difficult at first, my daily life has remained largely unaffected because the brain and stereo vision compensate the anomaly.

But more important than the actual event is the idea that this incident could have been avoided. Don't let it happen to you or a co-worker. Take time to assess safety conditions, and do it again in six months or a year; additional hazards arise in an ever-changing research environment. Safety deserves your thoughtful consideration, now, before YOUR accident. If even one other injury can be prevented by publication of this accident account, then more positive than negative outcome may result from the mistake.

As in Case History No. 1, this victim stated that, "I was not wearing goggles." In this case, the reason stated was, "In order to see the 532-nm pump beam propagation . . ." In applications that require observing a specific wavelength, yet maintaining protection from a hazardous level of energy density, custom eyewear can be specified as will be explained later in this chapter. Another myth dispelled; but, again, proper laser eyewear management requires familiarity with available services in this critical field of laser protective eyewear.

CASE HISTORY NO. 3

This is an interesting case and was printed in the November 1981 issue of *Laser Focus.* *

A Naval Research Laboratory chemist who was struck in the eye with a laser beam this summer is still suffering from the injury. The victim, who requested his name not be printed, was hit by 585-nanometer dye laser light that had reflected off an angle-turned frequency doubler when he bent over to adjust a stepper-motor drive. Although his vision has gradually improved, he told *Laser Focus* he lost much of the high-resolution capability in the eye.

The chemist, who's worked with lasers for five years and considers himself a "laser jock", was "amazed" by how little laser energy it took to do so much harm. Measurements made after the accident showed that the pulse back-reflected off the frequency doubler carried only about 25 microjoules. But that was enough to punch a hole through multiple layers of eye tissue and to cause hemorrhaging. The result was a blood blister over the mucula lutea, the part of the eye that provides visual acuity and which is necessary for tasks such as reading. The pulse energy would have been much higher—close to two millijoules—if the NRL group had not earlier taken steps to suppress amplified spontaneous emission in the dye amplifier chain, the victim said.

A surprising—and unsettling—discovery after the accident was how little the doctors knew about laser eye injuries. According to the injured, even retinal specialists were often reduced to guessing during treatment.

Laser Focus, November 1981.

The NRL researcher said he never saw a flash when the laser beam struck the eye. "I bent over and all of a sudden I couldn't see," he recalled. *He wasn't wearing safety glasses at the time*, which he said was common practice in the lab. One reason was that the laser—a YAG-pumped dye system—was run by computer and seldom needed adjustments that required close eye proximity to the beam. Also, he pointed out, laser systems that simultaneously produce numerous beams at wavelengths from the ultraviolet to the infrared are difficult to guard against. Since no single pair of goggles will block out all the beams, many lab workers choose to wear none at all. And in a darkened laser room, glasses that protect the wearer from laser light also obscure vision enough to raise the possibility of other hazards, such as hitting your head or tripping over cable.

The injured chemist criticized laser manufacturers for their method of compliance with Bureau of Radiological Health safety rules. Lasers are built in such a way "that to use one you've usually got to partly disassemble it," he said. "Laser companies should design their product so that it can actually be adjusted and used while complying with BRH rules." He also had harsh words for the maker of the frequency doubler that reflected into his eye. "A $10 beam-stop on that doubler could have prevented this whole thing from happening," he said.

There are two myths contained in this article. First, the victim was surprised at, "how little laser energy it took to do so much harm." That myth is dispelled by the damage threshold data in Chapter 3 that indicates as little as 10 $\mu J/cm^2$ in the ocular focus region can cause damage on the retina if the pulse duration is short. (No pulse duration was given in the article.) The second myth in this case, according to the victim, is ". . . in a darkened laser room, glasses that protect the wearer from laser light also obscure vision enough to raise the possibility of other hazards, such as hitting your head or tripping over cable." Visual transmission of colored filters has improved vastly, and the filter can be tailored to each specific application as explained in the section on Schott Optical Glass Company filters. In addition, the myth that lasers must be used in a darkened room is dispelled by the fact that a darkened room is much more hazardous than any slight gain made in reducing the intensity of reflected beams. All laser laboratories should be well lighted to aid workers wearing filter glasses that reduce visibility.

The following case histories have been published elsewhere.*

---

*Bouldrey, Edwin E. *National Safety News*, October 1981.

CASE HISTORY NO. 4

On June 27, 1979, a 31-year-old man with five years' experience with lasers sustained a neodymium laser burn to the left eye, causing an immediate dense central scotoma (blind spot).

Circumstances of injury were as follows: The patient was working on a commercial laser production line adjusting a laser through an opening in its top; this location required him to lean over the laser during beam alignment. In this position *his safety glasses slid up*, allowing the laser beam to pass underneath. The beam was reportedly being reflected from a piece of test paper in a plastic bag, a normal procedure during beam alignment, when the patient heard a snap and saw a bright after-image which lasted for 20 minutes before fading into a central scotoma.

The laser was neodymium YAG with a wavelength of 1.06 μm and a power of 1,500 mW. Beam diameter was 500 μm with a pulse duration of 0.1 seconds and a rate of about 10 pulses per second.

CASE HISTORY NO. 5

On January 10, 1977, a 32-year-old man with nine years' experience with lasers was injured in the left eye by a neodymium YAG laser causing the immediate onset of photopsias, pain, and visual blur.

Circumstances of injury were as follows: A laser had been custom-designed and built by the patient and others for his company. During modifications, a special beam-turning prism had been added, which incidentally had produced known stray reflections. Plans to block these reflections had not been carried out. The patient, who *was not wearing safety glasses*, had just entered the room to help align the beam when he felt a 'pop' and a sudden pain that lasted a few moments and that was associated with many photopsias and floaters.

The laser was Q-switched neodymium YAG with wavelength of 1.06 μm and peak power of one mW (six mJ per pulse). Pulse duration was six nsec with 10 pulses per second. Beam diameter was estimated as 1.5 to 3 mm.

CASE HISTORY NO. 6

On December 9, 1979, a 27-year-old graduate student with three years' experience with lasers sustained a neodymium laser burn to his left eye, causing an immediate central scotoma.

Circumstances of injury were as follows: The patient had assembled an experimental laser and had removed a beam block while making adjustments. As he leaned over he saw a flash in his peripheral vision and instinctively turned his eye towards the flash. There was an immediate central scotoma and decrease in visual acuity. He *was not wearing safety glasses* at the time of injury.

The laser was pulsed neodymium YAG with a wavelength of 1.06 μm and a power of one mJ with 50 to 100 kW peak power. Beam diameter was 2.5 mm and duration was 20 nsec, pulsing at probably 10 pulses per second.

CASE HISTORY NO. 7

On April 15, 1971, a 31-year-old man with many years' experience with lasers was struck in the left eye with an argon laser beam, causing an immediate paracentral visual blur.

Circumstances of injury were as follows: *Without wearing his safety glasses*, the patient was inspecting a clear glass laser beam splitter for dust particles as part of the normal production line procedure. During this examination, laser power was accidentally turned on by another person, causing the beam to strike the patient's eye.

The laser was continuous wave argon with wavelengths of 4,880 and 5,145 Å. Power incident on the cornea was 70 mW and beam diameter was 1.4 mm. The exposure duration depended on the blink reflex of the patient, estimated as being 0.125 seconds.

CASE HISTORY NO. 8

On April 21, 1978, a 24-year-old man with five months' experience with lasers was struck in the right eye with an argon laser beam, causing an immediate visual blur and scotoma.

Circumstances of injury were as follows: The patient was working *without safety glasses* on a laser production line aligning the beam when a reflection from a Brewster window struck his eye, causing him temporarily to see a flash towards which he instinctively looked, bringing on the injury.

The laster was continuous wave argon with wavelengths of 4,880 and 5,145 Å. Power incident on the cornea was 70 mW and beam diameter estimated to be less than or equal to 25 mW. Beam diameter at the time of injury was probably less than one mm.

CASE HISTORY NO. 9

On August 27, 1979, a 34-year-old man with six years' experience with lasers was injured in the left eye with a rhodamine dye laser, causing an immediate visual blur and scotoma. The patient was unaware of any other episodes of laser eye injury.

Circumstances of injury were as follows: *Without wearing safety glasses* the patient was adjusting a laser which had been constructed at his place of employment. While he was attaching a piece of cardboard to the laser to block a known light leak, the beam struck his eye, causing him to see an orange flash followed by an immediate scotoma and blur.

The laser was rhodamine pulsed dye with a wavelength of approximately 5,920 to 5,940 Å Power was 0.2 mJ (20 kW power) with a beam diameter of six mm. Beam duration was 10 nsec with a rate of 10 pulses per second.

CASE HISTORY NO. 10

On February 13, 1979, a 35-year-old man with 16 years experience with lasers was struck in the right eye with a beam from a krypton ion laser, resulting in an immediate visual blur and scotoma.

Circumstances of injury were as follows: The patient was assembling a production line laser and, although *not wearing safety glasses*, he had removed a safety screen to align the beam. The beam was focused on a jet of ethylene glycol and dimethyl sulfoxide solution when a bubble in the stream deflected the beam towards the patient's eye, causing the injury.

The laser was continuous wave krypton ion with a wavelength of 6,471 and 6,742 Å and power of approximately five W. The portion of the beam reflected to the patient's eye is unknown. Beam diameter in the fluid was 30 μm.

All of these case histories involved accidents that were caused by human error in judging the potential hazard of laser light of wavelengths in the ocular focus range by laser users not wearing protective eyewear. These examples also point out the relatively low power density or energy density required to cause retinal damage. Knowledge of laser damage thresholds is invaluable. The age-old rule of safety first in the laboratory should prevail, especially the wearing of protective eyewear in laser environments. No laser user properly wearing spectacles or goggles has ever reported a laser eye injury! That is the purpose of this chapter, to provide laser personnel with information about protective eyewear so that not only filter materials are evaluated but also that comfort, convenience, good visibility, and cost effectiveness may be presented as important considerations in selecting currently available eyewear.

DISCUSSION OF FRAME DESIGN

Comfort should be of paramount importance in selecting the type of frame or holder of the lens material of any eyewear. It is especially important to individuals not accustomed to wearing eyewear regularly. In a laser environment that requires constant use of protective eyewear, workers are more inclined to wear lightweight spectacles than the heavy, cumbersome goggle type. As much consideration should be given to frame selection as to the filter material. For example, if the eyewear can be folded to fit into a pocket case, the eyewear will be much more available than a pair of heavy, cumbersome goggles stored on a shelf or in a drawer. The following myths in laser eyewear design persist even after two decades of laser eyewear technology and extensive advertising in various media.

Myth No. 1 in Laser Eyewear Frames

Side shields or peripheral protection of some type is required. Simple geometric optics show that the possibility of side entrance or back reflection from properly worn spectacles is very remote, especially by wavelengths in the ocular focus region. Comfort and peripheral vision by the laser worker is overwhelmingly preferred in the author's opinion. For analysis of this myth, the diagrammatic sketch shown in Figure 8.2 clearly illustrates the difficulty a laser beam would have in reaching the retinal surfaces of an eye protected by properly fitted, lightweight spectacles without side shields. Because the eye is not directed toward the beam and is predominantly normal to the lens, reflections, which reduce the originating beam intensity, cannot enter the pupil at a 90°, or normal, position. Low-angle beam entry into the space between the temple of the eye and the eyewear frame certainly cannot enter the pupil, and high-angle reflections are also excluded. It would take a positive action by the wearer

Figure 8.2 The often used simple diagram in the top portion of this figure is *not* an accurate portrayal of the analysis of reflection of laser light impinging on the interior surface of protective spectacles. Optical analysis of such reflections, as shown in the lower diagram, dispells the myth that side shields are required for eye protection. The following is a statement from Dr. Martin Piltch, a Ph.D. laser physicist with 15 years' experience at the Los Alamos National Laboratory. "Applying the law of reflection from absorbing glass surfaces and interpreting Brewster's law as applied to reflection of laser light from the interior of spectacle surfaces, taking into account (1) the natural physical protection of the eye socket, and (2) the average distance of pupil to lens as 12 mm, the possibility of a hazardous laser reflection from the back surface of a glass lens of a standard spectacle frame that could enter normal to the iris and focusable on the retina is very remote."

to "cheat" outside the protective frame to "peek" directly at a beam; but even then, the entry of the beam would be at an angle and the foveal region of the retina would not, in all probability, be exposed. An example of the reduced intensity of a light beam can be experienced by a simple experiment with the sun. If a person's back is turned toward the sun, and standard spectacle-type sunglasses without side interference are properly in place, the intentional reflection from the sun by movement of the eyes within the frame does not cause discomfort, even though we know that the direct sun (50-100 mW/cm^2) can cause retinal damage in a matter of seconds if stared at directly. This demonstrates the extensive reduction in light intensity from back reflection off the lens material and the inability of the reflection to enter the pupil at an angle that would result in focusing on the small foveal spot of the retina. Apparently a geometric analysis, using Brewster's law, the absorptive properties of materials, and the simple laws of reflection and refraction had not been conducted by the early ANSI Z136.1 Committee that framed the first standard issued in 1973. It was specified in that edition that all laser protective goggles had to be "light tight"!

This presentation of facts for counteracting the myth of side shield protection from reflected laser light does not, however, suggest that side shields have no function in a laser environment. For example, in certain surgical procedures where close visual attention must be given to the affected, or target, area, the physician or others in attendance may desire side view protection not only from stray radiation but from debris emanating from the tissue involved. Also, welding or other material-processing operations could produce debris in the form of vapor, or even UV or IR radiation of a secondary nature that could affect the corneal tissue so that side protection is recommended. Again (it is repeated for emphasis), of all the eye damage accidents reported, no eye damage occurred to anyone properly wearing eyewear and certainly no damage occurred to anyone from a back reflection.

If side protection is desired, it is recommended that frames such as that shown in Figure 8.3 be considered rather than the cumbersome goggle type. Notice that the same type of glass can be used on the sides. It would seem to be cruel and unusual punishment to require laser personnel to wear any eyewear that is not comfortable. It is human nature to reduce uncomfortable wearing time to a minimum and to take risks without the eyewear; whereas a lightweight spectacle, carried in a pocket case when not needed, becomes a convenient aid in promoting eye protection. At the Los Alamos National Laboratory, the laser user is given a choice of many frames, within the constraint of safety types, for use in laser environments after evaluation of the workplace and after discussing frame styles with a safety engineer. Other employers also consider comfort in selecting eyewear for their employees. Many units of the

Figure 8.3 Photograph of spectacle-style frame with glass side shields. Recommended if side shields are required.

military are also concerned about the comfort factor, particularly the Air Force, which often requires personnel to wear the spectacles under other headgear. Shown in Figure 8.4 is the ultimate in laser eyewear frames—a lightweight, safety-approved, and stylish product selected by many laser users for the most comfort.

Myth No. 2 in Laser Eyewear Frames

Any frame of polished metal or other smooth, shiny surface will reflect the laser light into the eye. Again, this myth can be eliminated by analysis. First, the direct beam on the protective lens will be more intense than a reflected beam from any surface; and this fact persists for materials close to the eye as well as distant specular, or mirrorlike, reflections. Second, the surfaces of practically all frames are not flat, but rather curved, or rounded, so that any reflection will be dispersed and the beam will be reduced in intensity before it approaches the absorbing material of the lens. These simple explanations should dispell this myth, and those responsible for selecting frames for laser-protective eyewear are encouraged to disregard the frame material and to concentrate on comfort and, of course, on the selection of the proper filter for the lens material that is discussed in detail in the following section.

Figure 8.4 Photograph of three of the many styles of frames available for laser-protective lenses.

CHARACTERISTICS OF SCHOTT OPTICAL GLASS
COMPANY FILTERS

In the literature of the Schott Optical Glass Company's booklet titled *Color Filter Glass*, this quotation is appropriate to introduce this section.

> Colored glasses have been made into ornaments and expensive vessels since 200 B.C. Their use achieved artistic perfection in the translucent glass pictures of medieval church windows. The requirements of modern times have caused colored glass to be used ever increasingly for technological and scientific purposes. The most important consideration is no longer its color, but instead, the quantitatively specified dependence of its transmission on spectral wavelength, the so-called "spectral transmittance." This quantity should be specified for each colored glass type, and deviations must be kept to a minimum. Otto Schott, who originated the development of modern optical glass, also gave the science of colored glass a decided impetus. With the aid of his contributions, considerable improvement in industrial production of colored glasses was made possible. The methods we use at the present time not only permit production of larger pieces and greater thicknesses of colored glass but also guarantee outstanding homogeneity. Homogeneous pieces of colored optical glass are now available in relatively large sizes and thicknesses. . . . Fundamentally, all colored glasses can be distributed into two main classes. In the first class, the coloration is caused by colored simple or complex ions in true solution. With the second class, the so-called temperature colored glasses, the coloring agents are segregated submicroscopic particles the optical effect of which depends not only on their composition but also, to a considerable extent, on their size. Usually these particles only obtain their effective size by a posterior temperature treatment of the almost colorless basic glass.

The use of Schott glass as protective eyewear apparently started with the blue glass, BG-18, which was inserted as lenses into frames manufactured by American Optical Company. The author's first contact with BG-18 was in 1972 when, as part of my duties, I accepted an assignment as Safety Officer for the Laser Research and Technology Division. At that time, the eyewear was being used with resistance because of opaque side shields. Goggle styles were avoided and risks were being taken in several labs using Nd:YAG lasers emitting at $1.06\,\mu$m. Although the BG-18 glass absorbed this wavelength with excellent optical density (discussed in detail later in this chapter), the original purpose of the BG-18 glass was to protect against the early favorite laser, the ruby laser, emitting at 694 nm. This glass absorbs all red light, including control indicators and warning lights, and reduces visibility by about 50%. One physicist suggested that a clear Schott glass, KG-3, that was being used as a filter for the 1.06 μm wavelength, be considered for laser safety spectacles. Also, since he required correction eyewear, he asked, "Why couldn't someone provide prescription eyewear from the KG-3?" A bid request was prepared and sent to all known eyewear manufacturers. Only one firm in Albuquerque, New Mexico, presented a proposal for a contract to provide protective eyewear for prescription correction specifications. The proposal was accepted, and the Los Alamos National Laboratory inaugurated a program that has resulted in lightweight spectacles from essentially all of the Schott Optical Glass Company's color filters so that plano and corrective lenses may be produced for protection from any single or combination of laser wavelengths. Included in this service are a special dual frame that permits selecting two color filters for dual wavelengths and a special frame that can accommodate the same color filter glass in side protection as that used in the front lenses. This versatility in the fabrication of color glass filters permits customizing laser protection designed to control the hazards of any laser environment. As discussed in ensuing chapters, the selection of eyewear now allows optimum visual transmission with adequate absorption of harmful radiation and, as emphasized before, comfort and convenience for the laser user.

However, before evaluating the various candidates for eye protection, a discussion of the absorptive (or filtering) property, *optical density*, is presented here. *Optical density* is the ability of a material to absorb the wavelength of interest, expressed in mathematical terms. The ANSI Z136.1 Standard gives a rigorous mathematical definition of optical density (OD) as "logarithm to the base ten of the reciprocal of the transmittance:

$$D_\lambda = -\log_{10}\tau_\lambda,$$

where τ is transmittance." A useful and simplified expression, from R. C. Weast* is:

$$OD = \log_{10}\frac{I_i}{I_t}$$

---

*R. C. Weast, *Handbook of Chemistry and Physics*, 52nd ed. 1972.

where I_i is the intensity of light incident on a filter (in any unit, such as J/cm^2), and I_t is the intensity of light transmitted through the filter (in the same units). This simplified expression can be used to explain the author's practical approach to determining the optical density of eyewear material for a specific application. For example, if a steady Nd:YAG laser is being used, the first concern, as with any laser being evaluated, is, "What is the damage threshold value of the eye for this laser?" According to Figure 3.1 (Chap. 3), if the laser is continuous wave (CW), the value for the 1.06 μm wavelength is 5×10^{-3} W/cm^2 incident on the cornea. Next, the skin damage value for that value for Caucasian skin is 28 W/cm^2. Because the criteria for maximum available exposure should be that which can cause skin damage on direct or specular reflection, this laser should not be operated above 28 W/cm^2 without enclosure or remote control. (The ANSI Z136.1 Standard does not guide the laser user in what laser intensity limits should be imposed for possible direct exposure; but, if the intensity of a beam is such that it exceeds that for skin damage, or eyewear lens damage, which are similar, it is the author's opinion that this is the criteria for beam enclosure and remote operation.) Calculation of the optical density (OD) of the laser then would be:

$$OD = \log_{10} \frac{I_i}{I_t} = \log_{10} \frac{28}{5 \times 10^{-3}}$$

$$= \log_{10} \frac{28}{.005} = \log_{10} (5600)$$

$$= \log_{10} (5.6 \times 10^3) = \underline{3.7}$$

Another example would be for a pulsed Nd:YAG laser. Most of the reported eye damage accidents were from this type of laser, mainly because the user was not aware that the instantaneous energy in a fast-pulsed beam is more hazardous than that from a CW laser and can thus cause significant retinal damage. Also, because the beam is invisible, awareness is naturally reduced. If the pulse duration of this laser is 30 psec, the eye damage value listed in Figure 3.1 is 9 $\mu J/cm^2$. If we follow the criteria of skin damage threshold for maximum available exposure, that value is not available, and extrapolation is required. Suppose the value of 2.2 J/cm^2 for 100 nsec pulsed Nd:YAG radiation is used. The calculation for optical density would be:

$$OD = \log_{10} \frac{2.2 \, J/cm^2}{9 \, \mu J/cm^2} = \log_{10} \frac{2.2}{9 \times 10^{-6}}$$

$$= \log_{10} \frac{2.2}{0.000009} = \log_{10} (244{,}444)$$

$$= \log_{10} (2.4 \times 10^5) = \underline{\underline{5.4}}$$

Because the skin damage threshold will be less than 2.2 J/cm^2 for a 30 psec pulse, the actual OD required will be less than the 5.4 value; but ODs of 5 are readily available for the 1.06 μm wavelength in clear glass, so that would be recommended.

It should be emphasized here that ODs above 6 are not in the best interest of the user. Higher values are often quoted for plastic eye-protection devices, but it is a result of the manufacturers' desire to maintain other properties of the material and not because the "darker" shades are necessary for protection. The plastics would not withstand beam intensities if such high values were of practical use. (See later section in this chapter on damage threshold values.) It behooves the plastics manufacturers to improve the visual transmission of the laser eyewear by reducing the optical densities from ridiculous numbers like 11, 14, 15 *and even* 16, claimed by one manufacturer, especially when luminous transmissions are as low as 20%. The luminous transmissions available for most wavelengths in the color glass are quite good; for example, the KG-3 and KG-5 Schott glasses have nearly 90% visual transmission, and at 3 mm have optical densities at 1.06 μm on the order of 5. This compares with a luminous transmission of 45% and an OD of 14 claimed for a plastic material manufactured by the Glendale Optical Company. Optical densities above 6 specified by users indicate that management of eye-protection information with regard to eyewear specifications needs to show more concern for the comfort of the user and yet provide adequate eye protection. What good is a material that has an optical density so high that it theoretically will protect eyes against a beam intensity that would destroy eyewear lenses? That is certainly not a practical approach to controlling a laser hazard and providing the laser user with light-weight, comfortable eyewear with good visibility—the objective we all should strive for.

ULTRAVIOLET PROTECTION

According to the *Encyclopaedia Britannica*:

ultraviolet light [is] that portion of the electromagnetic spectrum adjacent to the short wavelength, or violet end of the light range. Often called black light, ultraviolet light is invisible to the human eye; but, when it falls on certain surfaces, it causes them to fluoresce, or emit visible light. . . . The ultraviolet spectrum is usually divided into two regions: near ultraviolet (nearer the visible spectrum), with wavelengths 2000 to 3800 angstrom units (200 to 380 nm); and far ultraviolet, with wavelengths 100 to 2000 angstrom units (10 to 20 nm).

As discussed in Chapter 1, the ultraviolet portion of the spectrum, for all practical laser safety concerns, ranges from 10 to 350 nm (see Chapter 1, Fig. 1.3). Because many clear, or uncolored, materials such as plastic and glass absorb ultraviolet radiation to some extent, laser eye protection should not reduce visibility appreciably in a laser environment associated with ultraviolet laser radiation. However, to be certain that protective materials have a consistent and known spectral transmittance, it is recommended that the three materials described below be used in selecting laser-protective eyewear for UV radiation.

Schott Glass GG-9 is one suggested filter for UV laser-protective eyewear because of its transmittance shown in the curve of Figure 8.5. This glass has good visual transmission (nearly 90%), can be prepared to prescription specifications, and adequately protects with a minimum optical density of five below 380 nm with 3 mm thickness. This glass can be ground to correct specifications, can be worn as regular "street wear," and does not have to be replaced when working with the UV radiation.

Optical Radiation Company's UV400 clear plastic is an excellent absorber of UV as shown in the transmittance curve in Figure 8.6. The material can be used for corrective needs, coated with wuartz for scratch resistance, and used effectively in absorbing UV, especially eximer laser wavelengths.

Laser Shades, Inc. Laser-550 is described in a later section in this chapter, it is used for UV and short blue wavelengths.

DEVICES FOR THE OCULAR FOCUS REGION

The range of wavelengths in the region of the spectrum, known as the "ocular focus" region, requires the greatest care in selecting laser-protective eyewear. This range (0.35-1.40 μm) is also the "retinal hazard" region of the spectrum because only these wavelengths are transmitted through the outer ocular components and are absorbed by the interior, or retinal, components of the eye. As mentioned previously, these light rays entering normally to the cornea are focused onto the retina by a factor of 100,000 times. Consequently, laser beams emitting radiation in this range of wavelengths do not need to be of very high

Figure 8.5 Transmittance curve for Schott's glass GG-9.

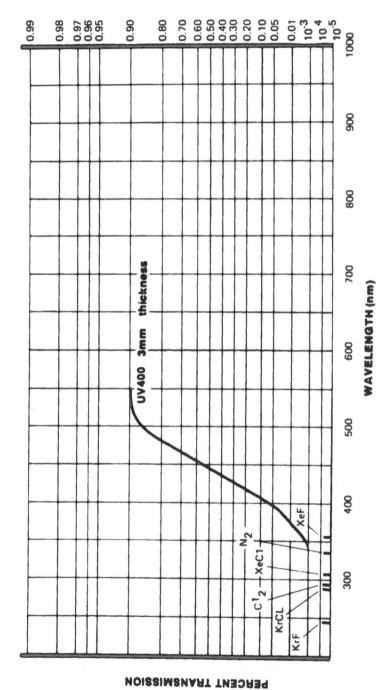

Figure 8.6   Transmittance curve for Optical Radiation Company's plastic UV-400.

intensity, especially pulsed laser beams. Thus, only limited laser radiation should be permitted in an environment that could expose laser personnel to pulsed laser beams.

A case history will illustrate the incredibly low intensity of ocular focus wavelength that could cause retinal damage from directly viewing the beam or from a specular reflection. In an experimental set-up at the Los Alamos National Laboratory using a fast-pulsed (30 psec) Nd:YAG laser, the small oscillator that originated the low-intensity beam before amplification to approximately 50 J (a very high energy level) was permitted to intermittently pulse every several seconds. At that time, before animal experiments were performed with a similar laser at Los Alamos, the relatively "weak" output of the oscillator was ignored until an experimentor was accidentally struck in the eye by the pulsed beam. He was adjusting a component in the beam's path near the oscillator output orifice, and the pulsed radiation reached the retina of the eye. The intensity of the beam entering the eye was estimated to be approximately 5 $\mu J/cm^2$. After several weeks of intermittent observation of the lesion by an ophthalmologist, it was determined that no permanent damage had resulted from the exposure. Later, experiments with a similar laser and a rhesus monkey's eye established the damage threshold at 9 $\mu J/cm^2$. A policy was soon established at the Laboratory that any pulsed laser emitting wavelengths in the ocular focus region was classified as Class 4, and appropriate controls were required to be instituted.

Chapter 3 describes the skin damage threshold values in the ocular focus wavelength range. The values listed range from 0.1 to 2.2 J/cm^2. It is recommended that personnel not be exposed to intensities above skin damage levels but that these lasers be enclosed or remotely operated. For example, the case history of the Los Alamos incident described above included corrective action to prevent further exposures. This action included insertion of a magnetic beam stop that was electronically connected to the laser controls. A key was required to activate the magnetic shutter, and the responsible operator initiated a countdown procedure before key insertion. Also, although personnel were permitted to be in the control area during the 50 J amplification, all wore protective eyewear and, in addition, turned their heads from the beam line and closed their eyes during the single 30 psec pulse. They were protected from the amplified beam by a physical barrier. This protective action is acceptable for single-pulsed lasers used in research activities where the scientists and technicians are knowledgeable; but, in general, the rule should be: if intensity levels are above skin damage levels, enclose the beam during operation of the laser. In any case, get approval of the Laser Safety Officer as explained in the ANSI Z136.1 Standard.

By using the criteria established above for maximum available radiation as that level of intensity above skin damage values, the maximum optical density for any eyewear selected for protection from ocular focusable wavelengths can be easily calculated as follows:

$$OD = \log_{10} \frac{I_i}{I_t} = \log_{10} \frac{2 \text{ J/cm}^2}{10 \, \mu\text{J/cm}^2\,*}$$

$$= \log_{10} \frac{10^\dagger}{10 \times 10^{-6}} = \log_{10} 10^6 = \underline{\underline{6}}$$

By using the value of 2 J/cm^2 for the skin damage value that would be incident on the eyewear because no values are available for 30 psec pulsed Nd:YAG (1.06 μm wavelength), a considerable (but unknown) safety factor is included. It has been established in eye and skin damage experiments that the shorter the duration of the pulse, the lower the damage value. So, in the above calculation two safety factors are known to exist by assumption of the person doing the computation. Despite these safety factors the optical density required for eye protection from the 30-psec, 10 J/cm^2 (maximum value permissible) Nd:YAG laser is 6. This value is typical for lasers of limited intensity permitted to be used when personnel are liable, at usual risk levels, to be exposed to above-skin-damage intensities of ocular focus wavelengths. This value of 6 for optical density of absorbing glass or plastic is a practical upper limit that permits optimum visual transmission and optimum lens thickness for eyewear. In addition, it is a practical limit for preventing saturation of the absorbing media and even for preventing damage to the lenses as well as the skin. Values above 6 for optical density specifications are not realistic and should always be questioned. Some manufacturers of eyewear list optical densities as high as 14, and this is deplorable. One wonders why these manufacturers do not reduce the OD and improve the visual transmission so that better viewing conditions are provided.

Laser protective eyewear manufactured from glass material is always the preferred absorbing media for the ocular focus range of wavelengths for several reasons:

---

*Lowest recorded retinal damage threshold value incident on cornea.

†For simplicity, assume a safety factor of 5.

1. Selection from a series of glasses with increasing increments of optical density permits adequate protection with a maximum visual transmission permitted for the specific wavelength of interest.
2. Thickness of the glass can be selected to accomplish item 1.
3. Lightweight, comfortable frames of the proper fit can be specified for each individual.
4. Corrections can be provided with the glass lenses, even bifocals.
5. Glass materials are more scratch resistant, thus cost effective.
6. Lightweight glass spectacles can be worn in a pocket case and be readily available to the user.

Perhaps the most important reason for recommending glass laser eyewear, if any one is paramount, is the last reason, item 6. No incidents of eye damage have been reported by laser personnel wearing protective eyewear, except in the case described in a later section in this chapter involving a plastic lens.

Shown in Figure 8.7 is a series of transmission curves for several Schott glasses that absorb wavelengths in the lower visible spectrum range. This region contains wavelengths of several popular lasers, particularly argon, krypton, doubled Nd:YAG, and tuneable dye lasers. The properties of the glasses can be explained from these curves as follows. The coordinates of the curves are transmission fractions (horizontal lines) and wavelengths (vertical lines). In selecting the optimum glass, the appropriate wavelength is identified. For example, assume that the argon spectral line of 0.51 μm is the emission of concern. If the intensity of the beam is at or below the value for skin damage (see Chap. 3) for a CW argon (assumed), which is ~1 W/cm^2, then the optical density for a 3 mm thick lens would be computed as follows:

$$OD = \log_{10} \frac{I_i}{I_t} = \log_{10} \frac{1 \text{ W/cm}^2}{10 \text{ mW/cm}^2}$$

$$= \log_{10} \frac{1}{10^{-3}} = \log_{10} 10^3 = \underline{\underline{3}}$$

By reading the horizontal coordinate at OD = 3, the transmittance being 10^{-3}, the intersection with the vertical 0.51 μm (510 nm) line is near to and slightly to the left of the glass curve identified as OG-530, indicating an OD slightly more than three. These glasses are called cut-off filters because, as the curves show, the steep curves cut off, or absorb, all wavelengths below that specific one used in the calculation of optical density. For example, in the case of the

Figure 8.7 Transmittance curves for Schott's OG and RG glass series.

argon spectral line of 0.51 μm wavelength, all wavelengths below the 0.51 μm are cut off, or filtered, with a minimum OD of three. These curves can also be adjusted for varying thickness of glass by moving the horizontal transmittance lines by the distances indicated on the thickness chart on the right side of the curves. Increased thickness makes the intersection with the curves lower and thus results in higher optical densities. The reverse is true of the glass thicknesses less than the circled thickness for which the curves are plotted. Notice at the upper portion of the curves that, because the curves are steep, a high transmittance factor is obtained to permit as much visible (0.4-0.7 μm) light as possible to get through the filter and help the individual see better. Well lighted work areas are recommended for personnel required to wear protective eyewear for visible wavelength laser radiation. A conflict of interest is obvious—you need the same wavelength, or near that, of the laser in order to have good visibility. So, a compromise results. A minimum optical density for adequate protection is obtained, allowing as much visible light as possible to be transmitted.

Another important aspect of these curves, which are considerably more accurate than some plastic eyewear manufacturers provide, is the ability to select a glass of desired thickness to permit visual detection of the beam from a diffuse reflection yet protect the eye from damage. For example, in the case of the argon CW laser mentioned above, it is possible to calculate the thickness of the desired glass to permit this unusual benefit of protection selection. In all cases of eyewear evaluation, it is recommended that actual conditions of the filter applications be measured by instrumentation whenever possible and that laser users should *never intentionally view a direct beam.*

As would be expected from this set of curves, as the wavelength of interest increases, the cut-off feature of the series automatically cuts off more and more visible light and the visual transmission decreases. Visibility is good until the darker red glasses (RGs) are used; then only 15-20% visual transmittance is available. In cases where wavelengths above 0.55 μm are involved, the Schott series of blue glasses are recommended. Transmittance curves for several blue glasses are shown in Figure 8.8. Notice that BG-1 is a narrow-band filter rather than a cut-off type and is recommended for wavelengths in the rage 0.55-0.65 μm. This range is the most difficult to prescribe for laser eyewear because it is in the very center of the color sensitivity of the eye. For laser wavelengths in this region, care should be taken to use the minimum thickness necessary for protection so that the maximum transmission of visible light is permitted. The minimum thickness permitted for ophthalmologic quality is two millimeters, and even at this thickness a bright work environment is required for good visibility. As the wavelength of interest is increased beyond 0.65 μm, BG-18 is recommended. This glass, developed by Schott, is ideal for the ruby laser wave-

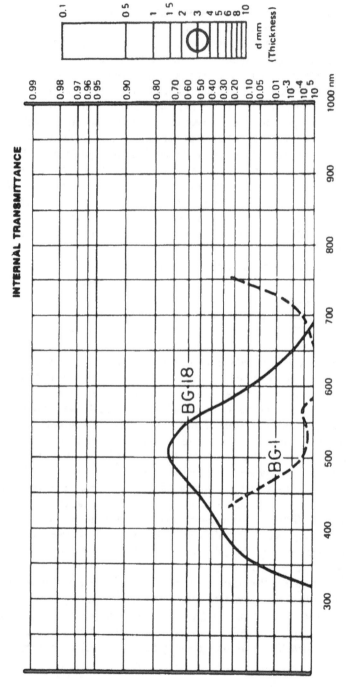

Figure 8.8 Transmittance curves for Schott's glasses BG-1 and BG-18.

length of 0.69 μm; but, as is the case of all blue glass made by Schott, not only is the red beam of the ruby laser absorbed, but all red lights on control panels, warning lights, etc. are also absorbed and are perceived as grey. BG-18 also absorbs the helium-neon laser wavelength of 0.63 μm, but generally these lasers are low-intensity emitters and eyewear is not required for those below the eye damage threshold of 10 mW/cm^2. Although the BG-18 is a good absorbing medium from 0.65 to 1.2 μm, it is not recommended for eyewear lenses above the visible wavelengths ($>$ 0.70 μm), because clear glasses (KGs) are available that possess high visual transmittance. These are discussed later in this chapter. The blue glasses BG-39 and BG-40, as shown in Figure 8.9, should be considered for special applications such as low-intensity dye lasers where visibility better than provided with BG-18 is possible.

Perhaps the most versatile of all Schott laser color filter glasses is KG-5 (improved KG-3). At 3 mm thickness, KG-5 absorbs, with excellent optical density, all laser wavelengths from 0.90 μm through 10 μm and beyond. As shown in Figure 8.1), this glass has nearly 90% visual transmittance, and therefore it is recommended for all wavelengths above 0.9 μm. It is particularly effective for protection against lasers emitting radiation in the invisible portion of the ocular focus region 0.9-1.4 μm. The very popular 1.06 μm wavelength provided by the Nd:YAG laser can be effectively absorbed by KG-5 yet permit high visibility in the workplace. In one application of welding with a Nd:YAG CW laser in which KG-3 was recommended as protective eyewear, the employees refused to believe that such a clear glass could prevent transmission of the laser beam. An experiment was suggested whereby the beam was intercepted by a power meter that indicated a beam intensity of 30 W/cm^2. As the employees watched with their eyelgasses in place, a lens of KG-3 was placed in front of the meter reducing the indicator to zero. The employees were impressed according to the supervising engineer. It was reported that the 3 mm lens cracked, without shattering, after approximately 10 sec of protection. This mode of failure is another adventage of using glass over plastics, which melt, bleach, and thin out (bubble), thus resulting in instantaneous loss of protection. The glass, by cracking. permits the aversion response of the eye to blink and close the eyelid thereby effecting eye protection.

It will be noted by comparing the transmittance curves of BG-18 and KG-5 that Nd:YAG at 1.06 μm is filtered by BG-18 at a higher optical density (OD = 11 claimed by Schott but at a value off scale in Fig. 8.8) than the OD = 5 for the KG-5. However, it should be obvious that the high visual transmittance of KG-5 is preferred and is recommended without exception for all wavelengths above 0.9 μm. If an OD of six is preferred, 4 mm lenses can be supplied but such a high optical density requirement should be questioned.

Figure 8.9 Transmittance curves for Schott's glasses BG-39 and BG-40.

Figure 8.10 Transmittance curve for Schott's glass KG-5.

The only known plastic material available for the 1.06 μm wavelength is manufactured by the Glendale Optical Company, but the visual transmittance is poor and it is not available in lightweight spectacles or in various thicknesses. Glendale does manufacture a Laser Gard Film, 0.005 in. thick, with good optical density, that could be used in some special applications. For example, in one vocational institute, the film was placed over a window between the instructor's office and the laser laboratory so that although the students were required to wear eye protection the office personnel were not. Often, this film can be engineered into laser-associated equipment and preclude the need for eyewear. Care should be taken to inspect the film regularly for bleaching or other defects that could transmit laser radiation.

INFRARED PROTECTION

Wavelengths longer than 1.4 μm do not require the same high optical densities as those in the ocular focus region because the radiation is not focused on the retoma—it is absorbed by the cornea. For example, if the same criteria discussed previously are applied to these long wavelengths for determining optical density (make the incident intensity, I_i, the value for skin damage threshold and make the transmitted intensity, I_t, that of the corneal damage threshold), the optical density for HF radiation of 2.7 μm, pulsed at 100 nsec, can be calculated as shown below (values from Chap. 3, Fig. 3.1).

$$OD = \log_{10} \frac{I_i}{I_t} = \log_{10} \frac{300 \times 10^{-3} \text{ J/cm}^2}{4 \times 10^{-3} \text{ J/cm}^2}$$

$$= \log_{10} \frac{300}{4} = \log_{10} 75 = \underline{\underline{1.9}}$$

A small safety factor could be applied if an optical density of two were specified. By using the transmittance curve for KG-5 in Figure 8.10 to determine the thickness required to obtain an OD of 2, the vertical line of 2.7 μm intersects the curve at 10^{-3} for the 3 mm thickness, indicating that that thickness provides an OD of 3. A measurement of the distance from 3 mm to 2 mm on the thickness chart, if applied vertically from the KG-5 curve upward, would not reach the 0.01 horizontal line, indicating that a 2 mm thickness would be adequate for an OD of 2. By this analysis, a lightweight spectacle can be provided to the laser user with confidence that adequate eye protection is available. It should be noted that all thickness specifications, unless otherwise specified, are considered the minimum in processing Schott glass filters so that a 3 mm thickness becomes 3.0-3.3 mm. Also, the minimum ophthalmologic lens thickness for corrective laser eyewear is 2.0 mm which becomes 2.0-2.2 mm after processing.

As shown in Figure 8.10, the KG-5 curve intersects the 10^{-5} transmittance line at 2.8 μm and disappears at an OD > 5. Almost all glass and plastic materials absorb infrared wavelengths above 3 or 4 μm. Especially at 10.6 =m (the radiation emitted by CO_2 lasers) essentially all materials absorb with high optical densities, even greater than 5 at a 3 mm thickness. This unique property of common materials makes fabrication of CO_2 lasing chamber output windows and amplifier windows expensive because special materials such as single-crystal salt (NaCl) must be prepared for this purpose. Consequently, so-called "streetwear" eyeglasses are recommended if correction is required because these spectacles are normally worn by the laser user and will be constantly available. There is no need to sacrifice visual transmission for infrared radiation above 4 μm wavelength or even lower if KG-5 glasses is available. One reason for using KG-5 for 10.6 μm radiation is that damage threshold values ($\sim$ 30 W/cm^2) are known for this glass whereas other materials are questionable.

Caution: Do not use eyewear of colored filters for CO_2 laser radiation. Clear, high-luminous transmittance glass and plastics are available at low cost and good quality. All regular (streetwear) lenses absorb the 10.6 μm wavelength with high, >5, optical densities.

Nd:YAG AND HARMONICS FILTERS

Shown in Figure 8.11 is a dual-lens holder frame that accommodates two color filter glasses for absorbing two ranges of wavelengths or two specific wavelengths simultaneously. This arrangement is particularly adaptable to the Nd:YAG laser and its harmonic emissions resulting from frequency doubling technicuqes. Because the frequency of the radiation is the reciprocal of the wavelength in the electromagnetic spectrum, if the frequency is doubled, the wavelength is halved. Consequently, if the Nd:YAG wavelength of 1.06 μm and its first harmonic wavelength of 0.53 μm are to be employed simultaneously, the base lens of KG-5 would absorb the 1.06 μm with an OD of 5; and the flip-down lens could be of OG-550 which absorbs the 0.53 μm also with an OD of 5, both lens thicknesses being 3 mm. It is possible to reduce the thickness of the OG-550 if the intensity of the secondary beam is low enough to permit a reduced optical density. By stating the intensities of the radiation needed to be absorbed, the thicknesses of the lenses can be minimized for reduced weight

Figure 8.11 Photograph of dual-lens holder frame, available without side shields. Notice optical density markings on left temple.

and improved comfort—the ultimate objective. Because the OG-550 is a cut-off filter with increasing otpical densities as the wavelength decreases, all lower Nd:YAG laser harmonic wavelengths, such as 0.27 μm, will also be absorbed. In addition, because the base KG-5 glass can be fabricated to accommodate correction, this combination of lenses is recommended for any Nd:YAG harmonic combination. A plastic goggle manufacturer, Glendale Optical Company, offers a goggle-style device that absorbs the 1.06 and 0.53 μm wavelengths with a reported OD of 4, but the luminous transmission is low, reportedly 45%; and the construction of the unit results in somewhat reduced vision, sealed enclosure that promotes condensation of perspiration, and bulkiness which makes wearing them uncomfortable. This type of soft plastic tends to scratch easily, cannot be contained in a convenient pocket case, and is relatively unsatisfactory. A third alternative method of eye protection, if comfort is not paramount but good luminous transmission is desired, would be to insert flat plates of KG-5 and OG-550 into the front fixture of a goggle frame design. Again, except for improved visibility, all the unsatisfactory properties of a goggle are incorporated in this alternative design.

HeNe AND Kr ALIGNMENT APPLICATIONS

Visual location of alignment beams at safe levels of intensity is often desired in some applications of laser technology. Usually the wavelengths are in the red color of the spectrum, either helium-neon (HeNe) or krypton (Kr) or even some dye wavelengths. Krypton lasers usually result from not being able to get high enough intensities of HeNe lasers, which are limited in maximum output power to 50 mW. Multiple watt Kr laser outputs are available. Alignment lasers often traverse multiple components in an optical path, therefore, the beam intensity decreases owing to absorption and diversion of the radiation and often is not discernible downstream. One application in which this situation was observed by the author was following the HeNe alignment beam path from the CO_2 double-discharge oscillator, through preamplifiers and amplifiers, reflected off several reflecting and focusing mirrors, and, finally, onto the target of a laser-fusion system. As the intensity of a 50 mW HeNe laser was too low to be observed by the time it reached the target chamber, a 2 W Kr laser was employed to obtain complete visual tracking of the beam. However, the intensity of the beam was above the eye damage threshold of 10 mW/cm in the upstream portion of the path so that eye protection was required. Visual detection of the beam from diffuse reflections was desired, so we chose a blue glass, BG-18, that would accommodate both requirements. The transmittance curve provided the information needed about optical density versus thickness at the 0.67 μm wavelength so that a controlled thickness was specified. The thickness chosen (2.2 mm) was checked by actual beam measurement using the blue color filter so that actual conditions were confirmed. The practice of testing adsorption of laser protective materials under actual operating conditions is always recommended. Schott glass samples of various thicknesses are available so that special applications of protective eyewear can be evaluated.

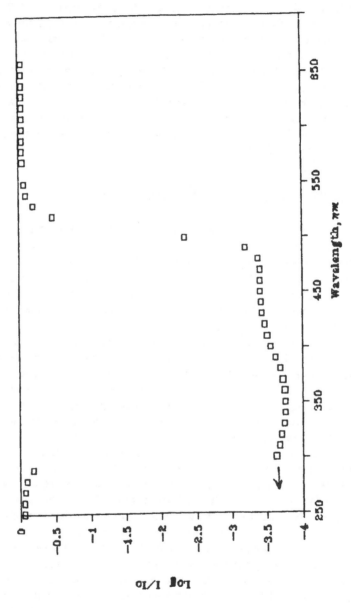

Figure 8.12   Transmission curve for Roman Thick Laser-550 lens.

The thickness recommended for use, after calculations by optical experts, should be such that corrective eyewear can be provided from prescription specifications.

SPECIAL CUT-OFF FILTER FOR UV AND BLUE WAVELENGTHS

The laser eyewear industry now has a rugged, lightweight isomer plastic lens that provides protection from laser wavelengths below 510 nm. The transmission curve for the material at 3 mm thickness is shown in Fig. 8.12 (courtesy Newport Corporation, Los Alamos Office). This curve was developed by a helium laser, shown in Fig. 8.13, and recorded by the operator shown at the

Figure 8.13 Photograph of technician obtaining an absorption (transmission) spectrum curve for a plastic material using a spectrophotometer. (Courtesy of Newport Corp.)

controls wearing the orange-colored lenses of the Laser-550 eyewear (provided by Laser Shades, Inc., Los Alamos, NM). Not only does this material absorb wavelengths in the 300-510 nm range at an optical density of 4 with a thickness of 3 mm, but it also provides high visual transmission. This ophthalmic quality plastic may be ground to prescription specifications including bifocals. Because of the high visual transmission, the lenses permit usage with good visibility in laboratory or manufacturing environments. Because the material absorbs the light blue as well as the UV wavelengths, the Laser-550 eyewear is an outstanding sun shade, providing protection from the harmful blue wavelengths (below 550 nm) and all the harmful ultraviolet rays. Visibility is so good that the eyewear is worn by night drivers to reduce the glare of oncoming headlights, yet not appreciably reducing visibility. No other sun shade absorbs so much of the spectrum at such optical density, yet provides such visibility. The material is especially suited for helium and ultraviolet laser wavelengths in the spectrum where biological eye damage thresholds are relatively unknown. The damage threshold values of the Laser-550 lenses are estimated to be higher than the thinner plastic lenses provided by the Glendale Optical Company's goggles, but lower than most 3-mm thick glass materials. These special lenses are available in any desired frames exclusively from Lase-R Shield, Inc. (Albuquerque, NM), and endorsed not only by laser users, but also pilots, skiers, sport shooters, and dentists (who use UV curing of plastic fillings), and recommended by physicians for patients with retinitis pigmentosa (retinal deterioration) and bilirubin (requiring exposure to UV for long periods, i.e., overnight). A hard coating is provided on the lense exteriors for scratch resistance, and the lenses are adaptable to the frame of choice. Lase-R Shield, Inc., the lens manufacturer, represents all domestic and several foreign frame manufacturers.

VISITORS' PROTECTION

Personnel in a laser environment assigned routine tasks that require eye protection should be provided with a style, design, frame, lenses, correction (if necessary), and other features of comfort as well as materials agreeable to the individual. The primary concern is to make everything convenient and suitable to the person involved so that the eyewear will be worn with comfort and confidence. Laser personnel should be indoctrinated with enough facts to be knowledgeable about the characteristics of the laser in use, the absorbing quality of the various lens materials, the availability of frames and styles of choice, and as much information as necessary to inform the worker of facts concerning the hazards involved and the part that protective eye devices play in controlling the hazards and protecting the eyes. Providing such information on appropriate eyewear for the environment should make it less likely that accidents involving laser damage to the eye will occur. It is rather obvious that lightweight, comfortable frames with the thinnest necessary lenses, having the highest possible luminous transmittance will be selected by people required to spend extensive periods of time in a laser environment. However, for those applications that require only occasional eye protection or that must accommodate spectators who visit the work place for short periods of time, the criteria for eyewear selection may differ.

Several options are available for providing occasional or short duration eye protection from laser light. Traditionally, the original goggle-type eyewear available from several sources has been selected. This line of so-called "anti-laser" goggles is handled by Lase-R Shield, Inc. as a manufacturer's representative. Over the years, Glendale Optical has developed three basic styles of laser eyewear. All have similar optical densities over the covered ranges of wavelengths. Apparently all viewing lenses have similar chemical composition. One of the goggle styles, and the one recommended, if spectacles are not chosen, contains a flat plate slot that can accommodate glass plates. Although the flexible vinyl body is large, bulky, and restricts peripheral vision, it does permit maximum visual transmission if Schott color filter glasses are used and the style has a double-frame option for single- or double-plate combinations. This style is shown in Figure 8.14. A second goggle style, shown in Figure 8.15, is of soft vinyl and lighter in weight than the glass-plate slotted type; the disadvantages are that all five available colors reduce luminous transmission and they are not scratch resistant. This latter unfavorable property does not make the plastic lenses cost effective. Because the goggle design cannot be placed in a pocket case as the spectacle can, personnel have a tendency to place the goggles on surfaces conducive to scratches and, probably most important, where they are not readily available for immediate use. Although the initial cost of the plastic goggle-type laser eyewear is 10-20% less than plano Schott glass lenses, the replacement costs do not make these devices cost effective over the preferred glass spectacles, especially for long-term use.

A third option of goggle-type laser-protective eyewear was recently developed to reduce bulk, tunnel vision, weight, and to replace the "clip-on" style (so called because it was designed to clip onto regular spectacle frames so that the colored plastic absorbing material of choice could be clipped on a suitable frame depending on the wavelength of interest). The Clip-Site series has been discontinued. The new style is shown in Figure 8.16 as a one-piece spectacle

Figure 8.14 Photograph of a variety of laser-protective devices available to laser users. The goggle styles that accommodate flat glass plates are located in the upper row at the extreme left and extreme right ends.

Figure 8.15 Photograph of soft vinyl plastic goggles manufactured by Glendale Optical Company for use with lasers emitting wavelengths in the ocular focus region of the spectrum.

Figure 8.16 Photograph of rigid-type plastic protective eyewear available for any specific wavelength from a variety of manufacturers.

design of rigid plastic that can be worn over regular corrective spectacles and is available in five colors to cover specific ranges of the spectrum. As with the soft vinyl goggle type, the disadvantages are reduced visual transmittance, increased weight for prescription requirements, scratch proneness, and they are not cost effective.

Some applications (for students or visitors) that require relatively short-term protection for laser personnel wearing corrective lenses (other than contacts) make goggle-type devices necessary; however, these styles should be used only after carefully considering all of the options.

Not to be considered for any laser-protective eyewear are devices that rely on materials coated on nonabsorbing substrates. Early in this emerging technology, evaluation of coated-type spectacles established that these devices were easily scratched, permitted radiation transmission, had directional absorbing properties, and had limited applications. (See discussion in later section of this chapter.)

LASE-R SHIELD, INC. SERVICES

Evolution of the current excellent choice of laser-protective eyewear containing a wide range of glass lenses has taken place over the past 15 years. In the early 1970s, use of the Schott blue glass filter, BG-18, was in vogue for absorbing the ruby laser wavelength of 0.69 μm; and, because of its wide absorption band that extended into the invisible portion of the ocular focus region, BG-18 was also employed for eye protection against Nd:YAG lasers emitting 1.06 μm radiation. However, blue glass had a luminous transmittance of approximately 45%; and, in addition, all red lights were absorbed. When the Laser Research and Technology Division at the Los Alamos National Laboratory was formed in 1972, extensive research was initiated with laser systems that emitted various wavelengths throughout the spectrum. The scientists assigned to research with the Nd:YAG wavelength of 1.06 μm utilized a "clear" Schott filter glass, designated as KG-3, for reducing intensity of the beam for detailed spectrum analysis. One of the physicists, Dr. Bob Watt, required corrective eyewear (bifocals) for routine laboratory work. His inquiry to the author (then designated as the Laser Safety Officer for the Division) to seek possible use of the KG-3 glass as corrective eyewear ultimately led to conversion of all eyewear in the Nd:YAG environments at Los Alamos to include the KG-3 glass filter.

The services available from Lase-R Shield, an ophthalmologic wholesale manufacturer, for laser-protective eyewear include evaluating the laser environment before recommending specific eyewear. Since this firm is a manufacturer's representative for a multitude of spectacle frames, safety goggles, and specialty eyewear, protective devices for any application can be provided. A supply of plano glass spectacles of the Schott series as well as plastic visitor goggles and spectacles are available for immediate delivery anywhere in the world.

If the eyewear is required as physical protection in some applications and conformance to the ANSI Z87.1-1979 is necessary, heat treatment of the laser glass filters by a tempering process can be provided. Most laser systems have extensive nonradiation hazards associated with their use, and this tempering service results in the need for only one pair of spectacles for routine laser environment employment. This applies to plano (uncorrected) wearers as well as those needing correction.

Prescription Lenses

Lase-R Shield offers the laser industry protective eyewear that contains individual corrections in a wide range of glass spectacle lenses. This unique feature permits laser users to combine the two functions of the lenses into one pair of spectacles. The prescription specifications are incorporated into the manufacturing of the lenses from any of the Schott glass filters. The minimum thickness of glass resulting from the correction grinding procedure is usually 3 mm. However, every effort is made to reduce the weight to a minimum so that it is possible to reduce further the thickness required for corrective lenses. The ultimate minimum thickness for manufactured eyewear is 2 mm. If analysis of the laser beam intensity versus the optical density of the glass results in adequate protection at 2 mm, this thickness, with tolerances of +0.2 mm and −0.0 mm, respectively would be recommended. Conversely, if the analysis indicated a thickness above 3 mm, this specification would be met to a practical maximum thickness of 4 mm. For example, a 4 mm thickness of KG-3 glass would provide an optical density of 6 at the Nd:YAG wavelength of $1.06\,\mu m$. Thicker lenses would prove to have excessive weight and therefore are not recommended. This latitude in custom servicing of specific applications is advantageous to the laser user.

Professional guidance is available from Lase-R Shield for eyewear fit and style, and permanent laser employees using corrective eyewear in laser environments are assured of satisfactory visual aid. All details of the ophthalmologic prescription are considered, including pupillary distance, frame width, nose rests, temple size, and single or bifocal correction. The bifocal construction can be the cemented-on-segment type or the Ben Franklin style. If the plastic UV-400 or Laser-550 lenses are desired for UV radiation, standard one-piece corrective lenses are available. These plastic lenses are coated with a thin film of quartz to provide a scratch-resistant surface for long-term wear. Pocket cases are provided for any type of spectacle. Special nose rests and temples, as well as elastic head bands, are also available. In addition to required inscription of optical density and wavelength data on eyewear, inscription of the owner's name is offered for prescription customers.

Special Applications and Services

By combining current laser-protective eyewear technology with essentially all hardware currently produced, Lase-R Shield has the resources to service any single wavelength or any combination of wavelengths that require eye protection. Tradeoffs are sometimes necessary to obtain adequate adsorption of the beam and yet provide good luminous transmission, or a choice of frames may be required to weigh comfort and convenience against optical density.

Tunable dye lasers

One specific example of a difficult choice in eyewear selection is the dye laser that is tuneable over a wide range of wavelengths in the ocular focus region. A conflict of desirable results exists whereby a good visual transmittance is necessary; but these are the same wavelengths (0.4-0.7 μm), that need to be absorbed for protection. Often, one narrow range in the emission possibilities is dominant in the use of the laser and that range would be the deciding factor in evaluating which Schott glass to use and in what thickness. Complementary lenses would then be determined to complete the coverage of all possible harmful rays. If the dominant wavelength range is between 0.55 and 0.65 μm (generally the requirement for tuneable dye lasers), for example, the blue glass of the Schott series designated BG-1 would be recommended at the minimum required thickness for adequate protection, because although it is a good absorber in this range, the luminous transmittance is low. Below the 0.55 μm wavelength, if the dye laser is tuneable that low, an OG series Schott glass would be recommended to adequately protect the eye with a minimum thickness; or a tradeoff of a lower visibility with increased absorption is possible. Wavelengths above 0.65 μm could be absorbed adequately by either the Schott glass identified as BG-18 or the glass BG-39 if better visual transmission is desired and lower optical density is acceptable. If correction is required by the laser user in an environment of such a range in wavelengths, and no narrow range of the spectrum dominates the use of the laser, a goggle-style frame that could accommodate several flat plates, as required one at a time, would be recommended. This would reduce the cost and make the eyewear available for anyone working in the environment. Comfort, in this instance, would have to be sacrificed because of the conflict of vision versus protection.

Schott Optical Glass Company glass color filter set

For those applications, particularly laser research over a wide portion of the spectrum, that could routinely use several 2 × 2 in. × 3 mm (variable) flat plates of the Schott series color filters, Lase-R Shield provides such a service for laser users. These filters could be used to determine which filter glass would absorb a particular wavelength for numerous reasons. One reason would be to determine the filter to specify for eyewear. Another reason would be to determine if a glass "window" could be used in a closed laser system if visibility of the target (such as welding) were necessary. Any combination of the glass squares, in various thicknesses, are available.

Laser safety consulting services

The selection of eyewear to protect ocular components from the damaging effect of laser beams is a complex process in many applications. A wide variety of spectacle frame and goggle styles containing various filtering materials is available to the laser user. Lase-R Shield, Inc. represents several manufacturers, in addition to its own line of plano and prescription lenses, attesting to the unique completeness of eye protection in this specialty line of eyewear. Many are in stock and ready for shipment. Their experience is unequaled in helping laser users evaluate the specific laser environment and then recommending a choice, generally, with consideration of style and comfort.

Discussion of laser-protective eyewear with an expert can often save buyers time and reduce the cost of protection from harmful radiation. For example, it is possible, as mentioned previously, to enclose a hazardous laser system except for a window (or port) or microscope eyepiece manufactured from a Schott glass or a vinyl film, to eliminate the need for eyewear. Also, one has to be assured of intellectual honesty when CO and CO_2 laser users inquire about appropriate eyewear for these wavelengths. (Any eyewear material absorbs these laser beams emitting above about 4 μm.) This type of consultation by Lase-R Shield's staff of experts is available without charge.

However, Lase-R Shield provides a consulting service, for a standard fee, for evaluation of laser environments. In addition to having expert eyewear specialists, they have consultants who are available for lectures, short courses, facility inspections, and manual preparation. For example, adherence to the ANSI Z136.1 "Safe Use of Lasers" should be required, especially with respect to (1) appropriate classification of the laser or laser system employed, (2) appropriate engineering controls, (3) complete medical surveillance, (4) personnel protection, and (5) control of all the nonlaser hazards associated with the technology. Often the consulting fee is less than the savings realized from the consultant's recommendations.

LASER-PROTECTIVE EYEWEAR: HOW SAFE IS IT?

A report of an eye injury in the January 1986 publication *Serious Accidents* by the U.S. Department of Energy (DOE) described how a laser scientist's retina was severely burned by a 1.06-μm repetitively pulsed laser. The researcher was wearing facility-approved plastic safety goggles.

The incident, which occurred at Lawrence Livermore National Laboratory (LLNL), Livermore, California, has raised questions about the practical damage thresholds for the various lens materials that are produced in the protective eye wear industry. For example, what is the relationship between optical density (OD) and protecting the eye from intensities that supposedly are absorbed by materials with manufacturer-quoted high ODs? Optical density is defined as the mathematical quantity represented by $\log_{10}$ of the ratio of the incident beam intensity to the transmitted beam intensity. High ODs do not guarantee protection of the eye from high intensity beams. Damage threshold values of the lens materials determine the maximum allowable intensity.

Another question concerns the American National Standards Institute's ANSI Z136.1 "Safe Use of Lasers" (ANSI Z136.1, 1986) and whether it contributes to lulling the laser user into a false sense of security by categorizing lasers as Class 4 and not offering more details about eye wear properties and damage-causing intensities.

A representative from LLNL stated that after the accident, a special projects group began to investigate the damage threshold values of several lens materials, but no experimental work has been completed and no results have been published yet. In fact, relatively little work has been done on laser damage effects in filter-glass and plastic lenses for current laser safety eye wear.

CASE STUDY: PLASTIC LENS DAMAGE AT LLNL

The following information appears in the January 1986 issue of *Serious Accidents.*

An LLNL laser researcher's right retina was severely burned by a repetitively pulsed Nd:YAG laser beam, which melted a hole through his LLNL-approved plastic laser safety goggles (see Figure 8.17). With full knowledge that the laser beam was on and passing through the cell that contained his experiment, the scientist looked directly into the cell. He said afterward that he had had confidence that the safety goggles would provide protection if he inadvertently looked into the beam.

Although the scientist had more than nine years of prior laser experience, most of it had been with single-pulse ultraviolet (UV) gas lasers. He believes his experience with UV lasers lessened his awareness of the hazards that were present; a UV beam hitting plastic goggles would have caused the plastic to fluoresce, giving a warning. In addition, he had not thought about the amount of average power that was present due to the rapidly pulsing laser. The 1.06-μm pulsed laser was running at a specific power of approximately 20 W/cm^2; the repetition rate was 10 Hz; the energy was approximately 0.2 J/pulse; and the length of each pulse was 30 ns.

The 1.06-μm laser beam presented a severe eye hazard (10^6 X Maximum Permissible Exposure) and a serious skin hazard (200 X Maximum Permissible Exposure). The injury was described as a 400-μm-diameter crater burned into the retina, surrounded by a 3-mm-diameter trauma region near the center of the field of vision. At this writing, it has not been determined whether the fovea—the region of the retina that absorbs 90% of the light for vision—was destroyed.

Figure 8.17 These plastic laser-safety goggles failed when an experimenter looked directly into a repetitively pulsed Nd:YAG laser beam. Although laser beams should not be viewed directly, and the goggle frames carried a warning against doing so, the consequences would have been less severe had he been wearing filter-glass goggles. (Courtesy U.S. Dept. of Energy, Washington, D.C.)

The project the scientist was working on was a typical tabletop experiment in which he was studying phase-conjugate back-reflection by stimulated Brillouin scattering in a cell containing sulfur hexafluoride gas. The setup was relatively simple with the laser beam focused into the cell. There were windows on both ends to allow passage of the beam and a beam stop beyond the exit window. Diagnostics were set up in front of the cell. The laser output was properly shuttered, confined to the experimental table, and significantly below eye level.

The scientist had been thinking about his day's work when it occurred to him that if laser-initiated gas breakdown were taking place within the cell, the results of his experiment would be affected. He looked into the cell for a symptom of gas breakdown, a spark.

Limitations of plastic-lens goggles

After the accident, several pairs of plastic-lens goggles similar to those the scientist was wearing were tested in the laser beam at the location where the victim's goggles had failed and at three other locations at higher specific powers. Nonprescription safety glasses made of a commercially available filter glass called KG-3 also were tested at the same locations. The plastic goggles were penetrated in 25 to 35 s with the 1.06-μm beam of light operating at a specific power of 16 W/cm^2. The filter-glass goggles survived the same beam operating at a specific power of 70 W/cm^2 for 5 min with no damage, although they were hot to the touch. Clearly there is a difference between filter-glass and plastic lenses in their capacities to handle high-average-power laser beams. The injury would probably not have occurred had the employee been wearing filter-glass eye protection.

Safety precautions

Lawrence Livermore recommends that laser filter-glass safety goggles be worn when working in laboratories where high-peak-power and high-average-power (repetitively pulsed or continuous wave) lasers are in operation. In addition, LLNL advises that plastic goggles not be worn by people who might be exposed to reflections or glints from such beams.

Furthermore, LLNL suggests that laser safety courses include the standards used for determining safe laser exposures and that job training include information about wavelengths and power levels of all lasers in the experimental area. Refresher training should be required periodically and whenever a person's work involves changing from one laser to another or from one experiment to another.

A periodic peer review of standard operating procedures for each laboratory is also recommended to confirm that all pertinent safety issues have been addressed.

CASE STUDY: BUCKBEE MEARS CO.

The high luminous transmission (85%) of KG-3 appears to the layman as a regular eyeglass material. It is difficult for the uninformed to realize the filterability of this special clear glass with respect to laser emissions in the infrared (IR) region of the spectrum.

In a demonstration experiment at Buckbee Mears Co., St. Paul, Minnesota, by Safety Manager Ray Grosman, a KG-3 lens from a pair of plano spectacles was positioned in the beam of a CW 1.06-μm wavelength Nd:YAG laser used in welding applications. After 20 s the glass was damaged, resulting in a typical fracture pattern for safety glass (See Figure 8.18). The output power was

Figure 8.18 This broken lens, which is made of the commercially available filter glass KG-3, is still intact without shattering after being irradiated for 20 s with a continuous-wave Nd:YAG laser. This result, realized by experiments at Buckbee Mears Co., St. Paul, Minnesota, is unlike the failure mode of the plastic lens in Figure 8.17, which shows penetration of the material. (Courtesy Fred Reed Optical Co., Albuquerque, N.M.)

recorded as 70 W as measured by a power meter for the 2-cm diameter beam. Thus, the power density on the lens was 22 W/cm^2.

The power meter, placed behind the lens during the exposure, maintained its zero calibration position, indicating complete absorption of the laser beam. Detection limit of the meter is estimated at 10 mW/cm^2.

Employees watching the demonstration were impressed by the results and agreed to wear KG-3 protective eye wear during the use of the laser.

CASE STUDY: LANL

In the early days of laser research and technology work (1972 to 1973) at Los Alamos National Laboratory (LANL), Los Alamos, New Mexico, lenses made of colored glass filters were determined to be preferable to plastic lenses for the laser wavelengths in the ocular focus region of the spectrum from 0.4 to 1.4 μm. The KG-3 IR-absorbing colored glass filter is essentially clear at visible wavelengths and yet adequately absorbs the 1.06-μm wavelength of the Nd: YAG laser. The higher inherent damage threshold values for such glass materials are favorable, also.

An experiment was set up in 1973 to test the lens made of a blue-colored commercial glass designated BG-18. It was developed for the ruby laser wavelength of 0.69 μm, but at that time it also was used for Nd:YAG's wavelength of 1.06 μm because of its high OD and because KG-3 was not being marketed at that time. The surface of the BG-18 lens showed no discernable damage after absorbing 5 J/cm^2 of energy from a 30-ps pulse of a 1.06-μm laser beam. The experiment did not include higher beam intensities.

The filter-glass materials became popular at LANL, not only because of their proven high ODs but also because eye corrections could be ground into the lenses, even bifocals, providing lightweight eyewear that could be worn with comfort for extended periods of time.

Following the information obtained from the LANL and Buckbee Mears experiments, KG-3 was used for protection against the 1.06-μm wavelength laser and other IR lasers because it is a cutoff filter for all IR. The high damage threshold value and the preferred mode of failure of the glass were important considerations.

Animal eye damage experiments at LANL were performed with a 30-ps single-pulsed Nd:YAG laser (Ham, 1974). The retinal eye damage threshold was determined as 10 μJ/cm^2 incident on the cornea. Thus there is no need for an OD above six in protective eye wear, for the Nd:YAG laser described because skin and probably eye wear lens damage would result at the level of intensity as shown by the following equation:

$$OD = \log_{10} \frac{10 \text{ J/cm}^2}{10 \text{ }\mu\text{J/cm}^2} = 6,$$

where 10 μJ/cm^2 is the estimated retinal damage value and 10 J/cm^2 is the estimated Caucasian skin and glass lens damage value for such a laser. Laser beams with intensities above 10 J/cm^2 should be enclosed unless the risk of direct or specular reflection is low. Consequently, OD values higher than six only encumber the wearer of protective lenses. KG-3 provides an OD of five at 3.3 mm thickness—the thickness used for regular corrective eyeglasses. Other color filter-glass applications are reported elsewhere.

To the author's knowledge, no eye damage has been reported by laser personnel wearing filter-glass lenses at LANL or any other research facilities during 15 years of intensive laser research and technology activities.

OTHER DAMAGE STUDIES

A report (1982) by H. Hack and N. Neuroth of Schott Glass Technologies Inc., Duryea, Pennsylvania, on the resistance of optical and colored glasses to 3-ns Nd:YAG laser pulses described an experiment using a "pulse of an Nd:YAG laser (1.060-μm wavelength, 3-ns pulse duration, output energy up to 1 J) generating damage at the surface and in the bulk optical glasses." The beam is described as slightly focused to exhibit a 2-mm diameter spot in the sample plane with a peak energy density of 50 J/cm^2. The surface damage was investigated on 5-mm-thick samples with a differential interference miscroscope.

In most cases, the damage observed was small pits probably caused by absorption centers in the polishing layer. Thresholds were more dependent on the polishing method than on the chemical composition of the glass. The data gathered showed typical surface-damage values ranging from 15 to 25 J/cm^2 for optical glasses and from 8 to 25 J/cm^2 for colored glasses like BG-18 and KG-3.

Internal damage was also studied with 20-mm thicknesses with damage thresholds ranging from 7 to more than 45 J/cm^2. Of course, bulk damage to gross cracking would require values higher than these reported for threshold damage. However, the bulk damage mechanism of cracking the glass lens of protective eye wear would permit blink response and thus avert further potential hazard. Also, the cracks would divert the beam and help reduce concentration by reflection. This is a different, but more desirable. damage mechanism than causing plastic materials to melt and permitting the laser beam to reach the eye without warning, as happened in the LLNL accident.

REFERENCES

ANSI Z136.1 "Safe Use of Lasers," American National Standards Institute, 1430 Broadway, New York, NY 10018 (1986).
Hack, H. and Neuroth, N. *Appl. Opt. 12*, 18 (1982).
Ham, W.T., Jr., et al., *Science 185*, 362 (1974).

9

Laser Personnel
Medical Surveillance

ANSI Z136.1 REQUIREMENTS

It is important that any laser user or any personnel assigned duties in a laser environment be aware of the potential hazards that radiation could pose to the eye, the organ most sensitive to the harmful rays of lasers. As in any potentially hazardous environment, these employees should also be informed that medical surveillance will be available to them because such hazards do exist in laser technology. As explained in Appendix E of the ANSI Z136.1 Standard:*

E1. Purpose of Medical Surveillance

The basic reasons for performing medical surveillance of personnel working in a laser environment are the same as for other potential health hazards. Medical surveillance examinations may include assessment of physical fitness to safely perform assigned duties, biological monitoring of exposure to a specific agent, and early detection of biologic damage or effect.

---

*Published by the Laser Institute of America, 12424 Research Parkway, Suite 130, Orlando, FL 32826.

Physical fitness assessments are used to determine whether an employee would be at increased or unusual risk in a particular environment. For workers using laser devices, the need for this type of assessment is most likely to be determined by factors other than laser radiation per se. Specific information on medical surveillance requirements that might exist because of other potential exposures, such as toxic gases, noise, ionizing radiation, etc, are outside the scope of this appendix.

Direct biological monitoring of laser radiation is impossible, and practical indirect monitoring through the use of personal dosimeters is not available.

Early detection of biologic change or damage presupposes that chronic or subacute effects may result from exposure to a particular agent at levels below that required to produce acute injury. Active intervention must then be possible to arrest further biological damage or to allow recovery from biological effects. Although chronic injury from laser radiation in the ultraviolet, near ultraviolet, blue portion of the visible, and near infrared regions appears to be theoretically possible, risks to workers using laser devices are primarily from accidental acute injuries. Based on risks involved with current uses of laser devices, medical surveillance requirements that should be incorporated into a formal standard appear to be minimal.

Other arguments in favor of performing extensive medical surveillance have been based on the fear that repeated accidents might occur and the workers would not report minimal acute injuries.

The low number of laser injuries that have been reported in the past 20 years and the excellent safety records with laser devices do not provide support to this argument.

E2. Medical Examinations

E2.1 Rationale for Examinations

E2.1.1 **Preassignment Medical Examinations.** Except for examination following suspected injury, these are the only examinations required by this standard. One purpose is to establish a baseline against which damage (primarily ocular) can be measured in the event of an accidental injury. A second purpose is to identify certain workers who might be at special risk from chronic exposure to selected continuous-wave lasers. For incidental workers (e.g., custodial, military personnel on maneuvers, clerical and supervisory personnel not working directly with lasers) only visual acuity measurement is required. For laser workers medical histories, visual acuity measurement, and selected examination protocols are required. The wavelength of laser radiation is the determinant of which specific protocols are required (see E2.2). Examinations should be performed by, or under the supervision of, an ophthalmologist or other qualified physician. Certain of the examination protocols may be performed by other qualified practitioners or technicians under the supervision of a physician. Many ophthalmologists may prefer to perform more thorough eye examinations to assess total visual function as opposed to limiting examinations to those areas that might be damaged by laser radiation at particular wavelengths. Some employers may find it advantageous to offer these more thorough examinations to their workers as a health benefit. For example, certain of the additional examinations, such as tonometry, may be of value in detecting unknown disease conditions, in this case glaucoma. Even though this type of problem is unrelated to work with lasers, appropriate medical intervention will promote a healthier work force. Although chronic skin damage from laser radiation has not been reported, and indeed seems unlikely, this area has not been adequately studied. Limited skin examinations are suggested to serve as a baseline until future epidemiologic study indicates whether they are needed or not.

E2.1.2 Periodic Medical Examinations. Periodic examinations are not required by this standard. At present no chronic health problems have been linked to working with lasers. Also, most uses of lasers do not result in chronic exposure of employees even to low levels of radiation. A large number of these examinations have been performed in the past, and no indication of any detectable biologic change was noted. Employers may wish to offer their employees periodic eye examinations or other medical examinations as a health benefit; however, there does not appear to be any valid reason to require such examinations as part of a medical surveillance program.

E2.1.3 Termination Medical Examinations. The primary purpose of termination examinations is for the legal protection of the employer against unwarranted claims for damage that might occur after an employee leaves a particular job. The decision on whether to offer or require such examinations is left to individual employers.

E2.2 Examination Protocols

E2.2.1 Ocular History. The past eye history and family history are reviewed. Any current complaints concerned with the eyes are noted.

Inquiry should be made into the general health status with a special emphasis upon systemic diseases which might produce ocular problems in regard to the performance cited in 6.1. The current refraction prescription and the date of the most recent examination should be recorded.

Certain medical conditions may cause the laser worker to be at an increased risk for chronic exposure. Use of photosensitizing medications, such as phenothiazines and psoralens, lower the threshold for biological effects in the skin, cornea, lens and retina of experimental animals exposed to ultraviolet and near ultraviolet radiation. Aphakic individuals would be subject to additional retinal exposure from near ultraviolet and ultraviolet radiation. Unless chronic viewing of these wavelengths is required, there should be no reason to deny employment of these individuals.

E2.2.2 Visual Acuity. Visual acuity for far and near vision should be measured with some standardized and reproducible method. Refraction corrections should be made if required for both distant and near test targets. If refractive corrections are not sufficient to change acuity to 20/20 (6/6) for distance, and Jaeger 1+ for near, a more extensive examination is indicated as defined in 6.3.

E2.2.3 Macular Function. An Amsler grid or similar pattern is used to test macular function for distortions and scotomas. The test should be administered in a fashion to minimize malingering and false negatives. If any distortions or missing portions of the grid pattern are present, the test is not normal.

E2.2.4 Contrast Sensitivity. Contrast (or glare) sensitivity should be documented by the Arden sine-wave patterns or similar acuity tests which include low contrast images.

E2.2.5 Examination of the Ocular Fundus with an Ophthalmoscope. This portion of the examination is to be administered to individuals whose ocular function in any of Sections E.2.2.1 through E.2.2.4 is not normal. The points to be covered are: the presence or absence of opacities in the media; the sharpness of outline of the optic disc; the color of the optic disc; the depth of the physiological cup, if present; the ratio of the size of the retinal veins to that of the retinal arteries; the presence or absence of a well-defined macula and the presence or absence of a foveal reflex; and any retinal pathology that can be seen with an ophthalmoscope (hyper-pigmentation, depigmentation, retinal degeneration, exudate, as well as any induced pathology associated with changes in macular function). Even small deviations from normal should be described and carefully localized.

E2.2.6 Skin Examination. Not required for replacement examinations of laser workers; however, suggested for employees with history of photosensitivity or working with ultraviolet lasers. Any previous dermatological abnormalities and family history are reviewed. Any current complaints concerned with the skin are noted as well as the history of medication usage, particularly concentrating on those drugs which are potentially photosensitizing.

Further examination should be based on the type of laser radiation, above the appropriate MPE levels, present in the individual's work environment.

E2.2.7 Other Examinations. Further examinations should be done as deemed necessary by the examiner.

E3. Medical Referral Following Suspected or Known Laser Injury

Any employee with a suspected eye injury should be referred to an ophthalmologist. Employees with skin injuries should be seen by a physician.

E4. Records and Record Retention

Complete and accurate records of all medical examinations (including specific test results) should be maintained for all personnel included in the medical surveillance program. Records should be retained for at least 30 years.

E5. Access To Records

The results of medical surveillance examinations should be discussed with the employee.

All non-personally identifiable records of the medical surveillance examinations acquired in E4 of these guidelines should be made available on written request to authorized physicians and medical consultants for epidemiological purposes. The record of individuals will, as is usual, be furnished upon request to their private physician.

E6. Epidemiologic Studies

Past use of lasers has generally been stringently controlled. Actual exposure of laser workers has been minimal or even nonexistent. It is not surprising that acute accidental injury has been rare and that the few reports of repeated eye examinations have not noted any chronic eye changes. For these reasons, the examination requirements of this standard are minimal. However, animal experiments with both laser and narrowband radiation indicate the potential for chronic damage from both subacute and chronic exposure to radiation at certain wavelengths. Lens opacities have been produced by radiation in the 0.295 to 0.45 μm range and are also theoretically possible from 0.75 to 1.4 μm.

Photochemical retinitis appears to be inducible by exposure to 0.35 to 0.5 μm radiation. If laser systems are developed that require chronic exposure of laser workers to even low levels of radiation at these wavelengths, it is recommended that such workers be included in the long-term epidemiologic studies and have periodic examinations of the appropriate eye structures.

Epidemiologic studies of workers with chronic skin exposure to laser radiation (particularly ultraviolet) are suggested.

E7. References

Friedman, A. I. The ophthalmic screening of laser workers. Ann Occup Hyg. 21: 277-279, 1978.

Hathaway, J. A., Stern, N., Soles, E. M., Leighton, E. Ocular medical surveillance on microwave and laser workers. J. Occup Med. 19: 683-688, 1977.

Hathaway, J. A. The Needs for Medical Surveillance of Laser and Microwave Workers. Current Concepts in Ergophthalmology. Societas Ergophthalmologica Internationalis. Sweden: 139-160; 1978.

Wolbarsht, M. L., and Landers, M. B. Testing visual capabilities for medical surveillance or to ensure job fitness. J. Occup Med. 27: 897-901; 1985.

In the 1973 edition of the ANSI Z136.1 Standard, the first standard issued for "Safe Use of Lasers," the medical surveillance of laser users was more rigid and more demanding that the current 1986 edition. Examinations were required every three years and essentially everyone in a laser environment, regardless of wavelength, was required to have an extensive protocol that required fundus photography of the retinal surfaces. Since only the ocular focus region of wavelengths, 0.35-1.40 μm, reach the retina, this requirement was changed to include only those personnel working in laser environments utilizing ocular focus wavelengths. Also, because fundus photography did not employ fast-speed film, bright illumination was necessary to provide adequate lighting to expose defects in the fundus area of the retina, resulting in discomfort and potential hazard during the photographic procedure. Physicists and physicians at the Los Alamos National Laboratory requested that the Standard be revised to eliminate the need for routine retinal fundus photography because, not only was the examination of the retina by an ophthalmoscope by the physician more revealing than a "hard photograph" of the fundus, but needless exposure to a bright light was eliminated. If an examination revealed defects in the retina that suggested photography, a neutral-density (clear) filter could be used in the fundus camera using fast-speed film that is now in common use and requires much less intense illumination. These changes are included in the 1986 edition of the Standard, Section 6, as quoted below. The *underlined* "shalls" are added by the author to emphasize a requirement rather than a recommendation.

6. Medical Surveillance

6.1 General. The rationale for medical surveillance requirements for personnel working in a laser environment and specific information of value to examining or attending physicians are included (above) in Appendix E. Medical surveillance requirements have been limited to those that are clearly indicated, based on known risks of particular kinds of laser radiation. Medical surveillance is required for personnel using other than Class 1, Class 2, Class 2a or Class 3a lasers and laser systems. Some employers may wish to provide their employees with additional examinations for medical-legal reasons, to conform with established principles of what constitutes a thorough ophthalmologic or dermatologic examination, or as part of a planned epidemiologic study. Further information is provided in Appendix E.

6.2 Personnel Categories. Each employee's category shall be determined by the LSO in charge of the installation involved. The individuals who should be under laser medical surveillance are defined in 6.2.1 and 6.2.2.

6.2.1 Incidental Personnel. Incidental personnel are those whose work makes it possible (but unlikely) that they will be exposed to laser energy sufficient to damage their eyes or skin, e.g., custodial, military personnel on maneuvers, clerical, and supervisory personnel not working directly with laser devices.

6.2.2 Laser Personnel. Laser personnel are those who work routinely in laser environments. These individuals are ordinarily fully protected by engineering controls or administrative procedures, or both.

6.3 General Procedures

6.3.1 Incidental personnel <u>shall</u> have an eye examination for visual acuity (see Appendix E for further details).

6.3.2 Laser personnel <u>shall</u> be subject to the following baseline eye examination:

Ocular history (E2.2.1). If the ocular history shows no problems and visual acuity (E2.2.2) is found to be 20/20 (6/6 in each eye for far, and Jaeger 1+ for near) with corrections (whether worn or not), central visual fields (E2.2.3) and contrast sensitivity (E2.2.4) are normal, no further examination is required. Laser workers with medical conditions noted in E2.2.1 should be evaluated carefully with respect to the potential for chronic exposure to laser radiation. Any deviations from acceptable performance will require an identification of the underlying pathology either by a funduscopic examination (E2.2.5), or other tests as determined appropriate by the responsible medical examiner.

6.4 Frequency of Medical Examinations. For both incidental and laser personnel, required examinations <u>shall</u> be performed prior to participation in laser work. Following any suspected laser injury, the pertinent required examinations will be repeated, in addition to whatever other examinations may be desired by the attending physician. Periodic examinations are not required.

MEDICAL SURVEILLANCE PROCEDURES AT THE LOS ALAMOS NATIONAL LABORATORY

The following description of the program at Los Alamos developed over the years by the Industrial Medicine, Industrial Safety, and Laser Technology groups appeared in the May 1984 issue of the *National Safety News*, written by the author with acknowledged help from several people.

Documented use of lasers at the Los Alamos National Laboratory dates back to 1964. Procedures were established by the Health Research Division to provide laser personnel with an examination of the eye by an ophthalmologist. Fundus photography of the posterior pole was performed after examination of the retinal surface with an opthalmoscope.

It was not until 1973 that the American National Standards Institute (ANSI) published the first laser-user standard, ANSI Z136.1, "Safe Use of Lasers"; and it contained a prescribed protocol for eye examinations for laser personnel.

The standard was revised in 1976 but retained all the medical surveillance requirements, including classification of personnel required to have retinal examinations. Those personnel exposed to any laser wavelengths, including those outside the ocular focus region (0.4 to 1.4 μm), required fundus photographs; and re-examination was necessary for "high risk" personnel every three years.

The standard again was revised in 1980, and several major changes were made in the Medical Surveillance Section.

It became apparent to the Occupational Medicine Group of the Health Research Division that the burgeoning use of lasers at the Laboratory in the late 1970's would require data processing of the information from ophthalmologic examinations and pertinent characteristics of the lasers in use as the preferred method of retaining permanent medical records. A computer program was designed to record information on memory tape and to permit recall of the pertinent information in sorting formats.

Details of the medical surveillance program included establishment of criteria for defining laser personnel, selecting the eye examination protocol required (based on laser characteristics), and processing of the data to provide a permanent record from which selected information could be recalled and printed on demand.

Definition of Laser Personnel. Personnel categories are defined in the 1980 edition of the ANSI Standard, Z136.1:

> Laser personnel are those who work routinely in laser environments. These individuals are ordinarily fully protected by engineering controls or administrative procedures, or both. Incidental personnel are those whose work makes it possible but unlikely that they are exposed to laser energy sufficient to damage their eyes or skin, i.e., custodial, clerical, and supervisory personnel not working directly with laser devices.

The assignment of the status category of an individual depends on the interpretation of the exposure risk involved by a designated laser safety officer, usually in concert with a representative of the Occupational Medicine Group.

Registration of Laser Personnel. The Laboratory's *Health, Safety, and Environment Manual*, Chapter 1, requires registration of all laser personnel by completion of the form shown in Figure 9.1. From this in-

─── HEALTH·SAFETY·ENVIRONMENT FORM ───

```
┌──────────────────────────────────┐
│    REGISTRATION FOR              │        DATE_____
│    LASER PERSONNEL               │
└──────────────────────────────────┘
```

LOS ALAMOS SCIENTIFIC LABORATORY

NOTE: Use separate form for each laser user.

All LASL laser personnel must be registered with H-Division. Completion of this form will enable Occupational Medicine (H–2) to determine and schedule appropriate eye examinations, and Safety (H–3) to review safety-related aspects of the laser environment. Laser personnel are defined in ANSI Z136.1 as "those who work routinely in laser environments, but who are ordinarily fully protected by safety features built into machines and procedures."

EMPLOYEE'S NAME _____GROUP_____

Z-NUMBER _____ MAIL STOP _____ PHONE_____

LOCATION OF LASER   TA- _____ BLDG. _____ROOM_____

LASER SPECIFICATIONS:	PULSED	REPETITIVELY PULSED	CONTINUOUS WAVE
TYPE	☐	☐	☐
LASING MEDIUM	_____	_____	_____
CLASSIFICATION*	_____	_____	_____
WAVELENGTH**	_____ μm	_____ μm	_____ μm
OUTPUT:			
power/energy	_____ J/pulse	_____ J/pulse	_____ W
radiant exposure/irradiance	_____ J/cm²	_____ W/cm²	_____ W/cm²
EMERGENT BEAM CHARACTERISTICS:			
diameter***	_____ cm	_____ cm	_____ cm
divergence	_____ mrad	_____ mrad	_____ mrad
PULSE CHARACTERISTICS:			
width	_____ s	_____ s	
repetition frequency (prf)****		_____ Hz	

*Class I, II, IIIa, IIIb, or IV

**Specify all wavelengths or wavelength range if using dye lasers, multiwavelength lasers, harmonic generation (frequency doubling, etc.), or Raman-shifted frequencies.

***If not circular, specify dimensions

****Lasers capable of operating at a prf of 1 Hz or more are considered repetitively pulsed lasers. Pulsed lasers cannot normally operate at a prf greater than 1 Hz.

GROUP LEADER SIGNATURE_____ GROUP _____ DATE_____

MAIL THIS FORM TO: SAFETY OFFICE, H–3, MS 403

Figure 9.1 Form for registration of laser users at Los Alamos National Laboratory.

formation an evaluation of the laser environment is made. The location of each laser is visited by an assigned Health Research Division Safety Engineer, and conditions are verified. Classification of personnel then is determined. Also, the type of eye examination protocol is recommended.

Determination of Eye Examination Protocol. The criteria for recommending an examination of the retina are:

- Whether the wavelength emanating from the laser being evaluated can be transmitted to the retina by the ocular system;
- Whether the power or energy density is sufficient to cause damage.

The nominal range in wavelength that can be transmitted, and focused with a gain in intensity of the order of 10^6 through the cornea, the lens, and the aqueous humor, and absorbed by the retinal tissue is 0.4 to 1.4 μm. This range of wavelength is called the "ocular focus or retinal hazard" region. All other wavelengths of lasers are absorbed by the outer components of the eye, particularly the cornea and lens.

Biological damage from pulsed laser radiation occurs at much lower energy density levels than from continuous wattage (cw) lasers, especially in the ocular focus range. The Laboratory requires that the protocol for the eye examination for all laser personnel using *any pulsed laser of any intensity or any cw laser with available power densities more than 50 mW/cm^2 and emitting wavelengths in the ocular focus region* include examination of the ocular fundus by an ophthalmologist. Laser personnel in all other laser environments require only a visual acuity test and examination of outer ocular components but not a retinal examination.

The original issue (1973) of ANSI Z136.1 Standard and the first revision (1976) called for re-examination of laser personnel every three years. However, based on the realization that the individual is the best "observer" of any appreciable change in retinal function, and because industry-wide experience has shown an extremely low eye injury rate, no re-examination requirement is contained in the 1980 edition of the ANSI Standard. Also, the revised standard does not require fundus photography—only examination by an ophthalmologist.

The ophthalmologist's examination includes more of the regina than a photograph, and a statement that the retina does not show lesions consistent with laser damage is more encompassing. The photographing of unusual lesions or suspicious areas therefore becomes the only photography done.

Processing Medical Surveillance Data. The Laser Personnel Registration Form is received by the Data Processing Section of the Occupational Medicine Group of the Health Research Division from the Technical Group employing a laser user. The pertinent information is entered onto the data base (description follows) by answering the questions of an interactive program. This program uses the social security number of the individual to create a "hash code," which then is used as the subscripts to file the record on a tree-structured data base. The record contains:

1. User's name;
2. Employee Z-number;
3. Employer;
4. Group (if Lab);
5. Mail stop;
6. Telephone number;
7. Date of last ophthalmologic exam;
8. Date started using laser;
9. Date stopped using laser;
10. Technical area (where laser is located);
11. Building number (where laser is located);
12. Room number (where laser is located);
13. Laser mode (continuous, pulsed, repetitively pulsed);
14. Wavelength (micrometers);
15. Energy or power density of laser (joules or watts per square centimeter); and
16. User type (continuous or incidental).

There are two types of user listing reports that are available from the data base. On one report all laser users can be listed with their locations, mail stop, telephone number, etc. Also listed is the date of the last ophthalmologic exam. This list is used to schedule users for ophthalmology exams and to locate them to inform them of the exam date and time.

The other listing report shows the users, the groups, and the laser characteristics. This list is used for laser surveys by various health groups.

A number of variations can be produced from these listings. Either list can be produced for specific groups, specific buildings, a special type of employee, a range of laser wavelengths, etc. Almost any desired listing can be produced from the data base, provided the information is contained in the record.

Figure 9.2 Form for ophthalmologist to use in retinal examination when required by the Los Alamos National Laboratory.

When a person is scheduled for an eye exam, an appointment is made with the ophthalmologist. The date and time of the exam are entered into the data base and a computer-written "reminder memo" is generated.

An OpScan mark-sensing device (reader) transmits and decodes the information from pencil marks on the forms to data that become recorded into a Medical Information System (MIS).

The MIS consists of a Digital Equipment Corporation PDP 11/40 Computer with 196,000 word core, supported by a DEC RP06, two RK05 disk drives, a seven-track tape drive, and a line printer.

The software is MUMPS v4-B that includes the operating system, the data base management system (DBMS), and the language. The system is accessible from 18 CRT terminals.

A separate form (see Figure 9.2), completed at the time of the examination, contains correction information for eyewear specification, if required. If a prescription is desired, a copy of the correction information is provided to the individual for submission with a laser eyewear purchase request. This form is kept as hard copy in the employee's medical record.

The procedures described above were in use when the author retired from the Los Alamos National Laboratory in 1982, and changes may have been made since then.

10
Critique of the ANSI Z136.1 Standard

The introduction of the latest edition (1986) of the American National Standard Institute's standard for the Safe Use of Lasers, identified as ANSI Z136.1 1986, states:

> An American National Standard implies a consensus of those substantially concerned with its scope and provisions. An American National Standard is intended as a guide to aid the manufacturer, the consumer, and the general public. The existence of an American National Standard does not in any respect preclude anyone, whether he has approved the standard or not, from manufacturing, marketing, purchasing, or using products, processes or procedures not conforming to the standard ...
> CAUTION NOTICE: This American National Standard may be revised or withdrawn at any time. The procedures of the American National Standards Institute require that action be taken to reaffirm, revise, or withdraw this standard no later than five years from the date of publication.

Previous editions were published in 1973 and 1980.

In the foreword of the Standard is the statement: "Suggestions for improvement of this standard will be welcome." That is the purpose of this chapter—to improve this Standard. A copy of this chapter will be sent to the Institute.

CLASSIFICATION OF LASERS

The basis for control of laser beams depend on the class assigned by the Standard to a particular laser. A set of tables provides for values of the beams' output, and a class is assigned to the laser depending on the output value of that particular laser.

The distinction between the four classes of lasers lies in the ability of the beam of a laser to cause biological (eye) damage from (a) direct or (b) diffusely reflected beams. *Class 1* lasers are "incapable of producing damaging radiation levels."

A *Class 2* laser emits in the visible portion of the spectrum (0.4 to 0.7 μm), and eye protection is normally afforded by the aversion response including the blink reflex. *Class 3* lasers (medium-power) are divided into two subclasses, 3a and 3b. A *Class 3* laser may be hazardous under direct and specular reflection viewing conditions, but the diffuse reflection is usually not a hazard. A *Class 3* laser is normally not a fire hazard. A *Class 4* laser (high-power) is a hazard to the eye and skin from the direct beam and sometimes from a diffuse reflection and also can be a fire hazard.

The classification table for *CW lasers* contains values in terms of power, or watts. However, the intensity of a laser beam is measured in terms of watts/ cm^2. The damage threshold values for CW lasers emitting in the ocular focus region (0.4-1.4 μm) of the spectrum, as well as the sun's rays, is on the order of 10 mW/cm^2 for one second incident on the cornea of the eye to cause damage to the retina (Winburn, 1985). It is recommended that Class 3 lasers be defined as those emitting in the ocular focus region (0.4-1.4) with beam intensities between 10 mW/cm^2 and 50 mW/cm^2. It is recommended that Class 4 lasers be defined as those emitting in the ocular focus region (0.4-1.4 μm) of the spectrum with an intensity greater than 50 mW/cm^2, the same power density used by the Los Alamos National Laboratory to stipulate Class 4 controls (Los Alamos National Laboratory Manual, 1987). Because the Standard does not stipulate power densities for laser classifications, the table should be revised to include consideration of the beam diameter in the classification of CW lasers. The philosophy behind this recommendation is that assigning Class 4 to lasers causing biological damage from diffuse reflections is an inexact requirement. Diffuse reflections cannot be determined accurately, varying from surface to surface, depending on reflectivity of the material and surface conditions. It appears more practical to specify the output of the laser using the "maximum available intensity" in terms of watts/cm^2. Any intensity downstream of the laser output coupler will then be less than that because of beam divergence and a safety factor is inherent in this approach.

The classification table for *pulsed* lasers contains values in terms of energy density, joules per square centimeters. Intensities for pulsed lasers and not for CW lasers shows inconsistancy. Data obtained from the majority of eye damage accidents with beams in the ocular focus region (0.4-1.4 μm) of the spectrum show that ultra-short-duration pulsed lasers of Nd:YAG wavelength, 1.06 μm, were the lasers involved (Winburn, 1985). Pulse durations of 1 nanosecond (10^{-9} s) to 1 picosecond (10^{-12} s) were the times of the bursts of the lasers, and intensity values on the order of microjoules per square centimeter incident on the cornea causing damage to the retina. These values have been documented as well as damage threshold values for a 30×10^{-12} s emission duration (Ham, 1974). The experimental data for these fast-pulsed lasers were presented to the ANSI Z136.1 Committee shortly after the information was published in 1974 by the investigative team headed by Dr. William Ham of the University of Virginia Commonwealth Medical School, a renowned biophysicist whose eye damage results have been used for other wavelengths of the single-pulsed laser classification table. It is recommended that the values reported by Ham, et al., be considered vital to the classification table, especially when so many accidents have been caused by these fast-pulsed Nd:YAG lasers. Also, it is recommended that any pulsed lasers in the ocular focus region of the spectrum be considered Class 4, because such low energy density can cause eye damage: 10×10^{-6} J/cm^2 with a 30 ps single pulse as reported by Ham.

NOMINAL HAZARD ZONE

The Standard contains the requirement that: "Where applicable, e.g., in the presence of unenclosed Class 3b and Class 4 beam paths, the LSO (Laser Safety Officer) shall be responsible for establishing the Nominal Hazard Zone (NHZ). The NHZ describes the space within which the level of direct, reflected or scattered radiation, during normal operation, exceeds the applicable MPE." This requirement is unnecessary, and the term NHZ should be eliminated from the Standard. It is confusing to safety officers and is causing extra intensity measurements from various surfaces surrounding the beam, which is ridiculous. If the laser involved is classified and the required controls for that class are in place, adequate precautions are assured. Why muddy the Standard with the questionable consideration of a "nominal hazard zone"? One can imagine multiple reflections through a window that would now be a hazard zone of infinitesimal scope!

DIFFRACTION-LIMITED SPOT SIZE

Before laser beams in the ocular focus region (0.4-1.4 μm) can be classified, the beam diameter must be evaluated to determine to what extent the ocular system of the eye can concentrate on the retina the intensity incident on the cornea. For example, if the beam diameter is 1 mm, the formula used to calculate the intensity on the retina should be applied as follows:

$$\text{Spot Size} = \frac{2.44 f \lambda}{d} \, ,$$

where f is focal length of eye, λ is wavelength, and d is beam diameter, so that if a 1 mm beam of HeNe (0.63 μm) is examined, the diffraction-limited spot size would be: 2.44 (17) (0.63 $\times$ 10^{-3}) = 26 μm (Winburn, *What Every Engineer Should Know About Lasers*, 1987). An explanation of this important aspect of focusing of the eye should be included in the definition section and discussed extensively so that non-technical users of the Standard understand the concept, and so that the intensity on the retina can be calculated and compared to the damage threshold values of approximately 1 W/cm^2 or 1 J/cm^2 (30 ps) for CW and pulsed lasers, respectively (Goldman, 1975). Also, only 80% of the power or energy is transmitted.

EYE PROTECTION

Evolution of protective eyewear for laser personnel has resulted in a need to revise this section of the Standard. The concept of optical density should be simplified and recent damage threshold values for eyewear lenses should be considered (Winburn, "Laser Protective Eyewear: How Safe Is It?", 1987). The optical density formula in the Standard is unnecessarily complicated:

$$D\lambda = \log_{10} \frac{Hp}{MPE} = -\log_{10} T\lambda \, ,$$

where Hp is expressed in the same units as the appropriate MPE and $T\lambda$ is the transmittance of the filter at the specific wavelength (see Section 8). A simpler representation of optical density is suggested.

$$OD = \log_{10} \frac{I_i}{I_t} \, ,$$

where I_i is the intensity incident on the lens and I_t is the intensity transmitted through the lens. If actual eye damage thresholds (Winburn, 1985) are used, instead of MPE values in the denominator, and the skin damage values are used for the numerator, the ODs are much more practical. (Intensities above the skin damage values should be enclosed.) Excessive optical densities can result by using the Standard's method, causing unnecessary increases in weight (thickness) and decreases in visual transmission, yet not protect the user because of low damage thresholds. It is recommended that the lowest reported values for damage to skin and eye be used rather than MPE's, which the Standard admits ". . . are below known hazardous levels."

Also, it is important that eyewear manufacturers be required to inscribe the lens material with its damage threshold value, not fictional high optical densities. What a sham to claim an exceptionally high optical density, but know the damage threshold value is very low in comparison. Damage threshold values for plastic and glass lenses have been reported recently (Winburn, "Laser Protective Eyewear: How Safe Is It?", 1987).

The need to have side shields for eyewear protection in the ocular focus region has been debunked (Winburn, 1985), so that eyewear need not be so cumberson. However, if the risk is high for exposure to Class 4 lasers emitting in the UV or IR regions, the wearer should consider side shields. These details should be explained in the Standard.

MEDICAL SURVEILLANCE

The Standard states, "medical surveillance requirements have been limited to those that are clearly indicated, based on known risks of particular kinds of radiation." However, the class and wavelength requirements are not clearly stated, and this Section, 6, should be clarified as follows (under Section 6.3.2):

> *Laser personnel* shall be subject to the following baseline eye examination for Class 3b and Class 4 laser environments:
>
> 1. if lasers are emitting wavelengths in the ocular focus region (0.35-1.40 μm) of the spectrum, the retinal surfaces of the eye shall be examined by an ophthalmologist and condition recorded; and
> 2. if lasers are emitting wavelengths outside the ocular focus region (0.35-1.40 μm) of the spectrum, only exterior surfaces of the eye need to be examined and the visual acuity test given.

The Standard needs to clarify this difference in examinations because UV and IR (outside retinal hazard zone) wavelengths do not reach the retina.

Also, the Reference list in the Appendix E7 should include this second edition of *Practical Laser Safety*. Part of a chapter covers Medical Surveillance as practiced by the Los Alamos National Laboratory that employs several hundred laser personnel. (This reference should also be listed in the Appendix to the Standard under C2, D7, F6.1, and G1.5. Many references are outdated.)

SUMMARY

As stated in the introduction to the Standard, it is essentially a guide to aid in the control of laser hazards. There are obvious flaws contained therein that need correction and constructive criticism is encouraged.

It is recommended that the following changes, deletions, or additions be made:

1. Classify lasers so that current Class 4 controls be applied to laser systems that can cause biological damage from direct or specular reflections; using all available threshold damage radiation levels, especially for the ultrafast Nd:YAG pulsed lasers. Footnotes may be used if experimental data is not corroborated, but list names, e.g., Dr. William Ham.

2. Classify CW lasers according to intensity, W/cm^2 not power, W.

3. Extract all references to and requirements for the Nominal Hazard Zone (NHZ).

4. Explain the effect of beam diameter on the ocular system resulting from the diffraction-limited spot size.

5. Require protective eyewear manufacturers to explain in their literature the damage threshold value for the lens material and have it enscribed on the lenses, keeping the optical density identification the same, using the intensity required to damage the lens physically in one second.

6. Differentiate the requirements for eye examinations for laser personnel exposed to wavelengths (a) in the ocular focus region, and (b) outside this range.

7. Update references in all appendices so that recent information published can be available to laser personnel.

8. Periodically, optical coating companies enter eyewear candidates for laser protection. The coatings reflect the radiation rather than absorb the harmful rays so that the high reflectivity of the coated surfaces does not attenuate the beam of a laser but, instead, causes the beam to seek another path. Manufacturers warn the users that the coated eyewear are not useful in laser environments where laser personnel are not wearing

coated lenses. It has been established that the reflective coatings' maximum protection is obtained when the beam is perpendicular to the reflective surface and that the protective factor decreases in proportion to the amount of angular change because the coated surface is actually a series of multiple coatings. If a compensating coat is applied to reduce this effect, the luminous transmission is also reduced. These optical coatings have a lower visibility factor than absorbing filters, especially the glass materials. Also, coatings are more fragile than the continguous absorbers, and breaks in the coat or scratches provide openings for radiation transmission. The substrates for the coatings are usually a glass material, and coefficients of expansion have to be precisely matched so that temperature variations (from tropical to frigid climates) do not cause spalling. All these factors tend to eliminate coatings not only because of physical characteristics but also from the high cost. The ANSI Z136.1 Standard should contain a paragraph on this subject, warning users of the hazards.

Emphasis on laser safety should be placed on a practical approach to protection of laser personnel. This can be accomplished by evaluating the intensity of the beam in relation to biological damage, and provide the necessary administrative and engineering controls, especially appropriate eyewear.

REFERENCES

Goldman, Arnold I. (1975) "An Ultrastructural Study of the Effect of 30ps Pulses of 1064nm Laser Radiation on the Rhesus Monkey Retina," Ph.D. thesis, Dept. of Biophysics, Medical College of Virginia, Health Sciences Division, Virginia Commonwealth University, Richmond, VA.

Ham, W.T., Jr., et al. (1974) *Science* 185, 362.

Los Alamos National Laboratory Manual, (1987) Chapter 1, "Health, Safety and Environment," HSE-Division, LANL, P.O. Box 1663, Los Alamos, NM 87544.

Winburn, D. C., (1985) *Practical Laser Safety*, Marcel Dekker, 270 Madison Ave., New York, NY 10016.

Winburn, D. C. (April 1987) "Laser Protective Eye Wear: How Safe Is It?" *Laser Focus/Electro-Optics*, p. 136.

Winburn, D. C. (1987). *What Every Engineer Should Know About Lasers*, Marcel Dekker, 270 Madison Ave., New York, NY 10016.

11

ANSI Z131.2, Laser Safety in Fiber Optics

This standard was developed separately from the general standard, Z136.1, because the application of laser radiation was unique and the type of laser emission in the communications industry required special attention. Only those sections of Z136.2-1988 are reproduced here, with permission of the American National Standards Institute, that include mandatory compliance. The complete standard may be obtained from its publisher, the Laser Institute of America, 12424 Research Parkway, Suite 130, Orlando, FL 32826. An introductory notice is included as well as a foreword by the Z136.2 Committee Chairman and Secretary.

NOTICE

The rationale for the use of this standard is as follows:

1. Determine the service group classification of the Optical Fiber Communication Systems (OFCS) based on the risk for injury.
2. Apply the appropriate controls as specified for each service group.

The accessible emission limit (AEL) for each service group is derived from the product of the maximum permissible irradiance (MPI) and the area of a specified limiting aperture. To take into account the possibility for viewing with optical aids, the distance between the end of an optically radiating fiber and the limiting aperture is different for the different services groups (see Table 11.2). Thus, the fundamental method for determining the service group is to divide the power measured through the appropriate aperture, located at a specified distance, by the area of the limiting aperture and compare this value with the appropriate MPI in Table 11.3.

To avoid the necessity of carrying out such measurements, analytical approximations for beam divergence and beam diameter have been provided as has the AEL corresponding to each service group. These expressions are *approximations* and, in keeping with the philosophy of safety standards, are conservative. while there is little ambiguity with the expressions for multimode fibers (the NA is usually specified by the optical fiber manufacturer), there may be some ambiguity with the expressions for the single-mode fibers. Because this standard is written in general terms for the largest audience, there may be some confusion as to which *diameter*, ω_0, is used in the expressions that pertain to single-mode fibers, i.e., mode-field diameter or core diameter. Since in many cases the core diameter is known, but not the mode-field diameter (and vice versa), the expressions for beam diameter, beam divergence and AEL are *approximations* that will yield conservative results no matter which diameter is substituted for ω_0. For example, if the mode-field diameter is used, an error as large as a factor of two times may exist at certain distances (the computed AEL may be one half the actual AEL). If the core diameter is used, the error in the computed AEL will be considerably smaller. If the mode-field diameter is accurately known, then, ω_0 can be replaced with $\sqrt{2}\,\omega_0$ in all equations for single-mode fibers. This will reduce the error in the computed AEL but will still yield conservative results.

FOREWORD

(This foreword is not part of American National Standard Z136.2-1988)

This American National Standard provides guidance for the safe use of optical fiber communications systems (OFCS) utilizing laser diode and/or light emitting diode (LED) sources by defining control measures for each of four service group classifications. Once an OFCS is assigned the appropriate service group classification, there should be no need to carry out tedious measurements or calculations to meet the provisions of the standard. This standard provides the information necessary to assign a service group to unclassified OFCS, or modified OFCS requiring reclassification.

During normal system operation OFCS are completely enclosed. There is no accessible emission and, therefore, no hazard. During service, however, there may be accessible emission. Consequently, each OFCS is assigned a service group classification based on potential hazard. Service group determination is based solely on output characteristics regardless of the type of source, i.e., laser diode or LED.

At present no known OFCS operate at wavelengths greater than 1.7 μm or at power levels greater than 10 mW. However, guidance is provided for longer wavelengths and higher power levels and is compatible with ANSI Z136.1.

This standard has been published as part of the American National Standard Z136 series. The basic document is American National Standard for the Safe Use of Lasers, ANSI Z136.1. In general, this standard may be used independently of ANSI Z136.1. Instances where additional guidance contained in ANSI Z136.1 is required are noted in the Z136.2 document.

While there is considerable compatibility among existing laser safety standards, some requirements differ among state, federal and international standards, particularly with respect to signs, symbols and control measures.

Suggestions for improvement of this standard will be welcome. They should be sent to the American National Standards Institute, Inc., 1430 Broadway, New York, N.Y. 10018.

This standard was processed and approved for submittal to ANSI by Accredited Standards Committee Z136 on the Safe Use of Lasers, whose scope covers protection against hazards associated with the use of lasers and optically radiating diodes. Committee approval of the standard does not necessarily imply that all members voted for its approval. At the time it approved this standard, the Z136 Committee had 64 members, representing 39 organizations or as individuals, and assigned to eleven subcommittees.

1. General

1.1 Scope. This standard provides guidance for the safe use, maintenance, and service of optical fiber communications systems (OFCS) utilizing laser diodes or light emitting diodes (LED) with average output powers of less than 500 mW and wavelengths between 0.4 μm and 1 mm. This standard applies only to systems where the radiant energy is confined to an optical fiber during intended use. For OFCS that do not meet these criteria, ANSI Z136.1 should be consulted. Note: At present no known OFCS operate at wavelengths above 1.7 μm nor at power levels above 10 mW. However, guidance is provided for longer wavelengths and higher power levels and is compatible with ANSI Z136.1.

1.2 Intended Use. This standard is intended to be used by manufacturers of OFCS and by installation and service personnel. It provides detailed safety information for systems where access is in uncontrolled areas, and where the source parameters are uncertain or not under control of the user.

1.3 Application. During normal system operation, OFCS are completely enclosed and there is no accessible emission. During service, i.e., when connectors are removed, there is a risk of exposure. Therefore, a practical means for accomplishing the purpose of this standard is first to classify the OFCS according to its potential hazard during service and then to specify appropriate controls for each classification. The classification scheme in this standard is derived from that in ANSI Z136.1. However, in order to avoid conflict with the definitions of the classes in ANSI Z136.1, a scheme using four service groups (SG), was devised: SG1, SG2, SG3a and SG3b. Since the parameters used to define the different SGs are emitted power and beam divergence, devices such as lens connectors are also covered provided the corresponding beam divergence can be determined.

1.3.1 Laser Classification. The classification of a laser or laser system is based on the ability of the optical beam to cause damage to the eye. Four different classes of lasers and laser systems are defined in ANSI Z136.1: Class 1, 2, 3 and 4. A Class 1 system is one that is considered to be incapable of producing potentially hazardous radiation levels during normal operation and maintenance and is, therefore, exempt from any control measures or other forms of surveillance. A Class 2 (or low-power) system applies only to lasers with visible emission (0.4 μm to 0.7 μm), which may be viewed directly only under carefully controlled exposure conditions. These must have a cautionary label affixed to the external surfaces of the device. A Class 3 (or medium power) system requires control measures to prevent exposure to the direct beam. A Class 4 system requires the use of controls which prevent exposure of the eye and skin to the direct or diffusely reflected optical beam. A Class 2, Class 3, or Class 4 laser or laser system contained in a protective housing and operated in a lower classification (Class 1, Class 2, or Class 3) shall require specific control measures to maintain the lower classification.

By the very nature of an OFCS, the optical radiation is confined to the fiber core and therefore is inaccessible to the user. Thus, under normal operating conditions, the OFCS is inherently a Class 1 system. However, when an optical connector is removed during service and an optical instrument, such as an eye-loupe, hand magnifier or a microscope (i.e., optically-aided viewing), is used to view the end of an energized fiber, it is conceivable that all of the radiant energy can impinge on the eye. For this reason, the new service group hazard classification scheme, which considers the special viewing conditions when cables are disconnected, has been devised.

OFCS certified for a specific class by a manufacturer in accordance with the Federal Laser Product Performance Standard (Title 21 Code of Federal Regulations Part 1040) shall be considered as fulfilling the classification requirements of this standard under conditions of normal operation. In instances where the laser or laser system classification is not provided; where the OFCS is being serviced or installed; where an LED is used; where the class level may change because of a change from the use intended by the manufacturer or where engineering control measures are added or deleted (see Section 4), the OFCS shall be assigned the appropriate SG in accordance with the procedures described below.

1.3.2 Service Group Classification (OFCS). All OFCS normally operate as Class 1, and the only risk of exposure would occur during installation and service. Also, the emission from an optical fiber is usually divergent, and therefore the risk for injury is different from that of a conventional laser with the same operating parameters, i.e., power and wavelength. Thus, the basic classification scheme in ANSI Z136.1 has been augmented by the establishment of classification in terms of service groups. It must be recognized that this scheme relates specifically to the OFCS product and its potential hazard, based on operating characteristics. However, the conditions under which the OFCS is used and serviced, the level of safety training of individuals servicing the system, and other environmental and personnel factors are important considerations in determining the extent of safety control measures. Because such situations require informed judgements by responsible persons, major responsibility for such judgements should be assigned to a qualified person, e.g., a Laser Safety Officer (LSO).

Since all OFCS are designed to normally operate Class 1, and the only risk for exposure would occur during installation and service, the OFCS has been assigned one of four service groups, specifically: SG1, SG2, SG3a, and SG3b. An OFCS is classified SG1 if the total output power is less than the Accessible Emission Limit (AEL) for Class 1 and there is no risk of exceeding the Maximum Permissible Irradiance (MPI) when viewing the end of a fiber with a microscope, an eye-loupe or with the unaided eye. The limiting aperture and the AEL measurement distance for SG1 are 5 mm and 2.5 cm (the focal length of a 10x eye-loupe), respectively, for wavelengths between 400 nm and 1400 nm. Different measurement criteria are employed at longer wavelengths. An OFCS is classified SG2 if it emits at wavelengths between 400 nm and 700 nm and if it is potentially hazardous when it is viewed for more than 0.25 s. An OFCS is classified SG3a if it is hazardous when viewed with a microscope or an eye-loupe but is not hazardous when viewed with the unaided eye. An OFCS which meets none of the above criteria is classified SG3b, provided that the total power is less than 0.5 W. Table 11.1 summarizes the potential risk associated with each service group.

1.3.3 Procedure for Using this Standard. The recommended stepwise procedure for using this standard is as follows:

1.3.3.1 Determine the appropriate service group of the system using the manufacturer's specifications. If this information is not available, the measurement methods described in Section 9, and the hazard evaluation and classification descriptions given in Section 3, should be used.

Table 11.1 Potential Risk Associated with Each Service Group

OFCS Service Group	Hazard from viewing with the unaided, unprotected eye	Hazard from viewing with optical aides
SG1	No	No
SG2	**	**
SG3a	No	Yes
SG3b	Yes	Yes

**A hazard only for exposure durations exceeding 0.25 seconds

For quantitative evaluation of the potential hazard associated with a given OFCS, Sections 8 and 9 should be consulted. To use the MPI values in Section 8, first determine whether or not the possibility for a hazardous diffuse reflection exists, in which case the MPI values for an extended source shall be used.

1.3.3.2 Comply with the control measures specified for the appropriate service group, using the table below as a guide.

Service Group	Control Measures[a]	Medical Surveillance (Service Personnel)
SG1	Not applicable	Not applicable
SG2	Applicable	Not applicable
SG3a	Applicable	Not applicable
SG3b	Applicable	Applicable

[a]For installation and service.

1.4 OFCS Utilizing Laser Diodes. Current OFCS that incorporate lasers use Class 1 or Class 3b laser diode transmitters. In normal operation these systems are totally enclosed and fully shielded by protective devices, i.e., the fiber sheath and connectors. Under these conditions there is no accessible laser emission and hence no hazard to safety or health. For this reason these systems are classified Class 1. The optical fiber cables which interconnect various components in the systems, however, can be deliberately disconnected or accidentally broken and, under some circumstances, permit human access to optical emissions. All components such as optical connectors or associated test equipment whose removal would allow access to levels of optical radiation greater than the accessible emission limits of SG1 shall be identified and labeled with the appropriate label (Section 4) commensurate with the service group of the system.

In certain instances, installation and service procedures may permit direct access to the emission from a Class 3b laser diode. Normally, emission from a Class 3b laser, at wavelengths between 0.4 and 1.4 μm, can cause damage to the retina of the eye. The emission pattern of a diode laser (or LED) is highly divergent, unlike that of a conventional laser, unless lens connectors are used. This means that the beam irradiance (hence any risk for eye injury) decreases rapidly with distance from the

output connector. Inadvertent viewing of a disconnected or unterminated energized fiber with the unaided eye at distances greater than 10
cm from the connector normally will not cause eye injury. However,
damage may be possible under optically-aided viewing conditions or
when lens connectors are used.

Note: Although viewing a disconnected energized optical fiber with
the unaided eye at distances greater than a few cm from the connector
may not cause injury to the eye, this procedure should be avoided since
it can result in developing a potentially dangerous habit. Consequently,
verifying continuity of optical fibers by shining a flashlight into one end
and viewing the other end is not recommended. The recommended method for verifying continuity is to use the appropriate optical test set. Viewing the end of a fiber that is suspected to be energized should only be
done with an indirect image converter or similar device.

1.5 OFCS Utilizing LEDs. OFCS that utilize LED sources generally
emit incoherent energy at wavelengths greater than 700 nm and power
levels considerably less than laser diode sources. Since the potential risk
of injury is determined by the output characteristics of the OFCS, i.e.,
power, coherence, wavelength and beam divergence, it is, therefore, generally less for an LED system than that for a similar system utilizing laser
diodes. However, for purposes of uniformity, all control measures com-
mesurate with a specific service group apply to both laser diode systems
and LED systems.

1.6 Optical Fiber Test Sets. There are at present several different categories of test sets in use with OFCS, e.g., optical loss test set (OLTS)
optical time domain reflectometer (OTDR), optical bandwidth test set.
Although some OTDRs use a Class 1 laser, many use Class 3b lasers, as
do typical OLTSs. An OFCS test set is itself considered an OFCS and,
thus, all control measures in Section 4 apply to its use.

3. Hazard Evaluation and Classification

3.1 General. Three principal aspects of an OFCS application influence
the total hazard evaluation and, thereby, influence the application of
control measures:

1. The capability of the radiant energy to injure personnel
2. The environment in which the OFCS is used
3. The personnel who may use or be exposed to the radiant energy

The OFCS service group classification scheme is based on aspect 1. An OFCS classified in accordance with this standard shall be labeled with the appropriate service group level, see 4.2.3. In addition, the classification labeling specified in the Federal Laser Products Performance Standard should be used to satisfy the labeling requirement of 21 CFR 1040.10. If the laser has been modified, the guidance in 3.2.3 shall be used.

Aspects 2 and 3 vary with each OFCS application and cannot be readily standardized. The total hazard evaluation procedure shall consider all three aspects, although in most cases only aspect 1 influences the control measures that are applicable.

3.2 Laser and LED Considerations. The individual responsible for service group classification shall ensure that optical fiber output data, including data sufficient for determining the divergence of a connectorized fiber, are valid and in accordance with Section 9 (Measurements). Classification shall be based on the maximum output power available.

3.2.1 Multiwavelength Optical Fiber Communications Systems. The classification of OFCS capable of emitting numerous wavelengths shall be based on the most hazardous possible operation.

3.2.1.1 A multiwavelength OFCS, which by design can operate only as a single-wavelength system, shall be classified as a single-wavelength OFCS.

3.2.1.2 A multiwavelength OFCS, which by design can operate over two or more wavelength regions, shall require classification in each region of operation. The appropriate control measures for each region shall be taken.

3.2.2 Repetitively Pulsed Optical Fiber Communications Systems. The evaluation of OFCS with pulse-repetition frequencies (PRFs) less than 15 kHz requires the determination of the number of pulses occurring within one exposure duration and the MPI per pulse as given in Section 8. MPI values for continuous wave (CW) OFCS apply only for PRFs greater than 15 kHz.

3.2.3 Radiometric Parameters Required for Determining Service Group Classification.

3.2.3.1 Classification of essentially all OFCS into the appropriate service group requires the following parameters:

1. Wavelength(s) or wavelength range
2. For CW or repetitively pulsed OFCS: average power output, limiting exposure duration in a period of 100 s inherent in the design or intended use of the OFCS; and, depending on the optical fiber used, either the numerical aperture (NA) of a multimode fiber or the mode field (core) diameter (ω_0) of a single-mode fiber. When lens connectors are used, the beam diameter and beam divergence associated with the connector must be known

Table 11.2 Limiting Aperture and Measurement Distance for Establishing AELs[a]

OFCS Service Group	Limiting Aperture (mm)	Measurement Distance (cm)	Limiting Exposure Duration (s)
$0.4\ \mu m \leqslant \lambda \leqslant 0.7\ \mu m$			
SG1	5	2.5	100
SG2	5	2.5	0.25
SG3a	5	10	100
$0.7\ \mu m < \lambda \leqslant 1.4\ \mu m$			
SG1	5	2.5	100
SG2	N/A	N/A	N/A
SG3a	5	10	100
$1.4\ \mu m < \lambda \leqslant 1\ mm$			
SG1	1	2.5	10
SG2	N/A	N/A	N/A
SG3a	1	10	10

NOTE: The AEL for SG3b is determined by the average power at the connector.

[a]These values are based on anticipated practical viewing conditions and are different from the limiting apertures and limiting exposure durations found in ANSI Z136.1.

Table 11.3 MPI for Direct Ocular Exposure, Intrabeam Viewing, for OFCS

Wavelength, λ (μm)	Exposure Duration t (s)	Maximum Permissible (MPI)	Notes for Calculation and Measurement
Visible and Near Infrared			
0.400 to 0.700	10^{-9} to 1.8×10^{-5}	$1 \times 10^{-}$	
0.400 to 0.700	1.8×10^{-5} to 10	$3.6\, t^{3/4} \times 10^{-}$	
0.400 to 0.550	10 to 10^4	$20 \times 10^{-}$	
0.550 to 0.700	10 to T_1	$3.6\, t^{3/4} \times 10^{-}$	-mm limiting aperture
0.550 to 0.700	T_1 to 10^4	$20\, C_B \times 10^{-}$	ee 8.2 for multiple
0.400 to 0.700	10^4 to 3×10^4	$2\, C_B \times 10^{-}$	ulse limitations
0.700 to 1.050	10^{-9} to 1.8×10^{-5}	$10\, C_A \times 10^{-}$	
0.700 to 1.050	1.8×10^{-5} to 10^3	$3.6\, C_A\, t^{3/4} \times 10^{-}$	
1.051 to 1.400	10^{-9} to 5×10^{-5}	$1 \times 10^{-}$	
1.051 to 1.400	5×10^{-5} to 10^3	$1.8\, t^{3/4} \times 10^{-}$	
0.700 to 1.400	10^3 to 3×10^4	$640\, C_A \times 10^{-}$	
Far Infrared			
1.4 to 10^3	10^{-9} to 10^{-7}	10^{-}	-mm limiting aperture[a]
	10^{-7} to 10	$0.56\, t^{?}$	
	>10	0.	
1.54 only	10^{-9} to 10^{-6}	1.	

NOTES: $C_A = 1$ for λ = 0.400 to 0.700 μm

       $C_A = 10^2(\lambda - 0.700)$ for λ = 0.700 to 1.050 μm

       $C_A = 5$ for λ = 1.050 to 1.400 μm

       $C_B = 1$ for λ = 0.400 to 0.550 μm

       $C_B = 10^{15}(\lambda - 0.550)$ for λ = 0.550 to 0.700 μm

       $T_1 = 10 \times 10^{20}(\lambda - 0.550)$ for λ = 0.550 to 0.700 μm

[a]The limiting aperture for wavelengths between 0.1 and 1 mm is 11 mm.

3.2.3.2 Classification of extended-source OFCS requires that, in addition to the parameters listed in 3.2.3.1, the fiber source radiance or integrated radiance and the maximum viewing angular subtense (apparent visual angle) be known.

3.2.3.3 To determine the potential of an OFCS for producing injury, it is necessary to consider not only the output irradiance or radiant exposure, but also whether a hazard would exist if the total output were concentrated within the limiting aperture for the applicable wavelength range when optical instruments are intended to be used or to be present (see Table 11.3).

These terms are defined in two different ways, depending on whether the output is considered a point source or an extended source.

3.2.3.3.1 Most OFCS can be considered point sources, i.e., intrabeam viewing conditions apply. For such OFCS, the AELs are each the product of two factors, namely:

1. The intrabeam MPI for the eye (see Table 11.3) for the limiting exposure duration
2. The area of the limiting aperture for the MPI (see Table 11.2) expressed in cm^2

3.2.3.3.2 For extended-source OFCS that emit in the spectral range 0.4-1.4 μm, the AELs are determined by the power or energy output such that a theoretically perfect optical-viewing system (50 mm limiting entrance aperture, 5-mm exit aperture) would collect the entire OFCS output.

Note: If this approach is difficult to apply, the definition in 3.2.3.3.1 may be applied and will result in a conservative AEL.

3.3 OFCS Hazard Classification Definitions. The OFCS service group shall be determined according to its accessible emission level when operated at the maximum power for its intended use.

3.3.1 OFCS SG1.

3.3.1.1 Any OFCS containing an LED or laser that cannot emit accessible levels of optical radiation in excess of the SG1 AEL for an emission duration of 100 s, is designated SG1 and is exempt from all control measures and from any form of surveillance. The exemption strictly applies to optical radiation hazards and not to other potential hazards (see Section 7, Special Considerations).

3.3.1.2 OFCS intended for a specific use may be designated, by a qualified person, SG1 on the basis of that use for a classification duration less than 100 s, provided that the accessible emission level does not exceed the corresponding SG1 AELs for the maximum possible duration inherent in the design or intended use of the OFCS.

3.3.2 OFCS SG2.

3.3.2.1 Any OFCS containing an LED or laser emitting in the visible (0.4-0.7 μm) which can emit a total accessible radiant power exceeding the appropriate SG1 AEL for the maximum possible emission duration inherent in the design of the system, but not exceeding the SG1 AEL for an exposure duration of 0.25 s or less, is designated SG2.

3.3.3 OFCS SG3a.

3.3.3.1 OFCS designated SG3a includes:

1. Near-Infrared (0.7-1.4 μm) CW or repetitively pulsed OFCS which can emit accessible radiant power in excess of the SG1 AEL for an emission duration of 100 s but cannot emit an average radiant power in excess of the AEL for SG3a

2. Infrared (1.4 μm-1 mm) CW or repetitively pulsed OFCS which can emit accessible radiant power in excess of the SG1 AEL for an emission duration of 10 s, but cannot emit an average radiant power in excess of the AEL for SG3a

3. Visible (0.4-0.7 μm) CW or repetitively pulsed OFCS which can emit accessible radiant power in excess of the SG2 AEL for an emission duration greater than 0.25 s, but which cannot emit an average radiant power in excess of the AEL for SG3a

3.3.4 OFCS SG3b.

3.3.4.1 OFCS designated SG3b includes:

1. Near-infrared (0.7-1.4 μm) CW or repetitively pulsed OFCS which can emit accessible radiant power in excess of the SG3a AEL for an emission duration of 100 s, but which cannot emit an average radiant power in excess of 0.5 W

2. Infrared (1.4 μm-1 mm) CW or repetitively pulsed OFCS which can emit accessible radiant power in excess of the SG3a AEL for an emission duration of 10 s, but cannot emit an average radiant power in excess of 0.5 W

3. Visible (0.4-0.7 μm) CW or repetitively pulsed OFCS which can emit accessible radiant power in excess of the SG3a AEL for an emission duration of 100 s but which cannot emit an average radiant power in excess of 0.5 W

3.4   Environment in Which the Optical Fiber Communications System is Used. Following OFCS classification, environmental factors require consideration. Their importance in the total hazard evaluation depends on the OFCS classification. The decision to employ additional hazard controls, not specifically required in Section 4 of this standard, is influenced by environmental considerations principally for SG3b.

The probability of personnel exposure to hazardous OFCS radiation shall be considered and is influenced by whether a disconnected fiber (connector) of an OFCS is in a controlled or uncontrolled area.

Although exposure of unprotected personnel to a primary or specularly reflected beam of a conventional laser is a major concern, exposure to an OFCS beam is seldom of concern beyond 10 cm unless lens connectors are used. Therefore, calculations or measurements either of the irradiance or radiant exposure for the primary or specularly reflected beam at the location of bystanders is generally unnecessary.

Viewing the primary beam or a specular reflection from an OFCS with an optical instrument may present a hazard due to the instrument's light-gathering capability.

3.4.1   Optical Fiber Communications System Operations in an Uncontrolled Area. The step-by-step procedure described in 3.4.1.1 through 3.4.1.4 is recommended for evaluating SG3b OFCS in an uncontrolled area when necessary because of potential exposure of unprotected personnel.

3.4.1.1   Step 1. If the connector or emitting optical fiber is not mounted in a fixed position, determine the probable hazardous beam path(s).

3.4.1.2   Step 2. Determine extent of hazardous specular reflection, if any.

3.4.1.3   Step 3. Determine if extended-source criteria apply, e.g., for a multifiber connector, a cable or an array. Note, however, that the angular subtense of the source is normally sufficiently small at all practical viewing distances that intrabeam MPIs apply, except under magnification.

3.4.1.4 Step 4. Consider the likelihood of people being in the area of the OFCS beam, particularly when lens connectors are used.

3.5 Personnel. The personnel who may be in the vicinity of OFCS and its emitted beam(s), e.g., service and installation personnel, can influence the total hazard evaluation and, hence, influence the decision to adopt additional control measures not specifically required for the class of OFCS being employed.

3.5.1 If small children or others unable to read or understand warning labels may be exposed to hazardous OFCS radiation, the evaluation of the hazard is affected and control measures will require appropriate modification. (See Section 4.)

3.5.2 The type of personnel influences the total hazard evaluation. The following are considerations regarding personnel who may be exposed.

1. Maturity of judgment of the OFCS user(s)
2. General level of training and experience of OFCS user(s), e.g., high school students, military personnel, production line workers, service personnel
3. Awareness on the part of onlookers of potentially hazardous OFCS radiation, and of relevant safety precautions
4. Degree of training in OFCS safety of all individuals involved in the operation of the system
5. Reliability of individuals to follow standard operating procedures and recommended control procedures
6. Number of individuals and their location relative to the primary beam or reflections, and to potential for accidental exposure
7. Other hazards not due to OFCS radiation that may cause the individuals to react unexpectedly or which influence the choice of personnel protective equipment

4. Control Measures

4.1 General Considerations. Since OFCS are Class 1 laser products, an LSO is not required. However, when SG3b systems are serviced, management (employer) is responsible for the establishment and surveillance of the appropriate control measures.

Engineering control measures which are incorporated in, or added by the user to, the OFCS shall not be eliminated (notwithstanding their use not being required by this standard). Priority should be given to the use of engineering controls as an integral part of OFCS to control all applicable potential hazards.

If the engineering control measures of this section are impracticable or inadequate, protective equipment and administrative and procedural controls should be used. Administrative and procedural controls are methods or instructions which specify rules, or work practices, or both, which implement or supplement engineering controls and that may specify the use of personal protective equipment. The limits of any type of control measure, e.g., failure modes of cables and connectors or the inability of some personnel to read hazard warnings, should always be considered in developing an OFCS control program.

Note: Unless otherwise specified, administrative and procedural controls apply only to SG3a and SG3b OFCS.

4.2 Uncontrolled Area.   Under normal operating conditions OFCS are completely enclosed. The following controls are required to minimize potentially hazardous exposure:

4.2.1   When the service group classification for an OFCS has not been determined, safety connectors should be used in all uncontrolled areas.

4.2.2   Safety connectors shall be utilized with all SG3b OFCS located in uncontrolled areas.

4.2.3   For OFCS where the service group is known and cannot be changed to a higher service group anywhere within the OFCS, and where safety connectors are not used, labels which include the following wording shall be used:

For SG2 OFCS

CAUTION –
DISCONNECTED OPTICAL CONNECTORS
MAY EMIT OPTICAL RADIATION
DO NOT STARE INTO BEAM

For SG3a OFCS

CAUTION —
DISCONNECTED OPTICAL CONNECTORS
MAY EMIT OPTICAL RADIATION
DO NOT VIEW BEAM WITH
OPTICAL INSTRUMENTS

For SG3b OFCS

DANGER —
DISCONNECTED OPTICAL CONNECTORS
MAY EMIT OPTICAL RADIATION
AVOID DIRECT EYE EXPOSURE
TO THE BEAM

For all infrared systems, the words INVISIBLE OPTICAL RADIATION should be used in place of OPTICAL RADIATION. For all visible systems, i.e., for wavelengths greater than 400 nm and equal to or less than 700 nm, the words OPTICAL ENERGY may be substituted for the words OPTICAL RADIATION on all labels.

The labels shall be located in plain view on the equipment in the vicinity of the connectors, e.g., outlets, patch panels, patch cords.

4.2.4 If an optical fiber cable should accidently break or be removed from its normal position, the following practices should be followed:

1. Do not examine or stare into broken, severed or disconnected optical fibers or cables.
2. Arrange for a trained service person to repair or replace the optical fiber or cable.

4.3 Controlled Area. Under some OFCS installation or service conditions, e.g., splicing, removal of optical connectors, there may be accessible emission. Under these disconnect conditions the following administrative and procedural controls should be followed:

4.3.1 Only authorized train personnel shall be permitted to install or perform service on SG3b OFCS.

4.3.2 Only authorized trained personnel shall be permitted to use SG3b OFCS test equipment during installation or servicing.

4.3.3 All unauthorized personnel shall be excluded from the nominal hazard zone (NHZ) of SG3b OFCS during installation and service when there is a possibility that these systems may become energized.

4.3.4 During installation or service, a broken optical fiber or unterminated connector should only be viewed with an indirect image converter or with a filtered optical instrument of optical density (OD) sufficient to reduce the exposure to levels below the appropriate MPI, unless it has been verified that all optical transmitters are turned off.

4.3.5 During all splicing operations that require viewing the end of a fiber of a SG3b OFCS, laser diode or LED sources on the fiber involved shall be deenergized or viewing systems incorporating personal protection shall be employed.

4.3.6 Connectors which may emit SG3a or SG3b levels of optical radiation shall have affixed the warning label required by 4.2.3, unless a sign having the same wording is posted conspicuously nearby. A sign is normally used in an area with a large number of connectors, e.g., an equipment bay or cable closet.

4.3.7 The degree and level of education and training on safety concepts and procedures shall be in accordance with Section 5.

4.3.8 Eye protection devices which are specifically designed for protection against radiation from SG3b OFCS should be administratively required when engineering or other procedural administrative controls are inadequate to eliminate potential exposure in excess of the applicable MPI.

4.3.9 During service or installation of a SG3b OFCS where access panels or doors are removed or opened and the critical viewing distance could exceed 100 cm (as may occur when lens connectors are used), means shall be used to contain the beam to preclude exposure of nearby personnel.

5. Safety and Training Programs

5.1 Organizations. The management (employer) shall establish and maintain an adequate program for the control of OFCS hazards. Safety and training programs shall be instituted for personnel servicing and installing SG3a and SG3b OFCS. If a safety and training program is instituted, it shall be directed by individuals competent in the field of laser and OFCS safety. The program shall provide:

1. Background information on laser and LED OFCS
2. Safety information on the OFCS service groups
3. Guidance on the safe use of laser and LED OFCS and the adequacy of safety practices

5.2 Training. The training program shall enable technicians, installers, and service personnel to maintain a safe work environment. Service personnel shall be instructed periodically on the safety precautions, procedures and practices that are applicable to OFCS and on safety measures that will ensure, insofar as possible, that OFCS and any associated test equipment are properly operated and controlled to protect transient or uninformed personnel.

5.3 Responsibility of Individuals Working with OFCS. Each person engaged in the installation and service of OFCS is responsible for:

1. Observing all rules, procedures and practices established for the safe operation of OFCS
2. Notifying supervision immediately of conditions or practices that have the potential to cause personal injury or property damage
3. Reporting immediately to supervision any known or suspected abnormal exposure to optical radiation

6. Medical Surveillance of OFCS Personnel

6.1 Coverage.

6.1.1 Personnel assigned to work routinely on energized SG3b OFCS and whose job requires disconnecting optical connectors, using SG3b optical test equipment, e.g., loss test sets, optical time domain reflectometers, or performing splicing operations on SG3b OFCS shall be included in a medical surveillance program. These personnel shall have a baseline eye examination before starting work in an OFCS environment. The purposes of the medical surveillance program are:

1. To establish a baseline of ocular conditions against which any suspected damage can be measured in the event of accidental exposure to OFCS emission
2. To detect and document possible ocular damage as soon as possible

6.1.2 Employees who work in the general vicinity of OFCS but who are not routinely within the control distance of an emitting optical fiber should not be included in the medical surveillance program.

6.1.3 The rationale for medical surveillance requirements and specific information on recommended medical examinations are based on the guidelines provided in ANSI Z136.1, Appendix E. The eye examination is not designed to be part of a complete medical examination in which the total visual system is evaluated. The particular visual parameters selected for measurement examination are those deemed necessary for the particular job and/or those which may be affected by hazardous exposure to OFCS radiation.

6.3 Frequency of Medical Examinations. The examination shall be carried out prior to work with OFCS, immediately after a suspected abnormal exposure of the eye, or for specific eye complaints on an individual basis.

No periodic examinations are required, but may be offered in special circumstances when some special problem exists.

6.4 Records and Record Retention. Complete and accurate records of all medical examinations (including specific test results) shall be maintained for all personnel included in the medical surveillance program. Records should be retained for at least 30 years.

6.5 Access to Records. The results of medical surveillance examinations should be discussed with the employee.

All non personally identifiable records of the medical surveillance examinations acquired pursuant to 6.4 shall be made available on written request to authorized physicians and medical consultants for epidemiological purposes. All personally identifiable records upon request and with written permission of the employee shall be made available to a private physician or other designated medical personnel.

7. Special Considerations

7.1 Glass Particle Hazards. As well as the possible hazards to the human eye from optical radiation, there also is some risk of direct injury from small lengths or particles of fiber, particularly when cleaving fibers during splicing operations. All personnel should be warned of this hazard. The use of protective guards or shields should be considered, especially during cleaving or jointing operations.

Where practicable, discarded pieces of fiber should be collected in a suitable container to avoid subsequent embedding in clothing or skin.

7.2 Photocuring Hazards. Optical photocuring devices should be operated with caution to preclude exposure to UV and viewing of the light source. Protective filter lenses should be worn if direct viewing of the light source is necessary.

7.3 Solvents and Chemicals. Solvents and chemicals used for stripping fibers and cleaning and splicing fiber elements shall be used in adequately ventilated areas. The least hazardous solvents needed for a task should be used. Material Safety Data Sheets (MSDS) for each solvent or chemical should be consulted. Service personnel shall receive adequate safety instructions in the use of solvents and chemicals.

8. Criteria for Exposure of the Eye and Skin

MPI values are below known hazardous levels. Levels at the MPI values for visible wavelengths may be uncomfortable to view, thus, it is good practice to maintain exposure levels as far below the MPI values as is practicable.

When a laser emits radiation at several widely different wavelengths, or where pulses are superimposed on a CW background, computation of the MPI is complex. Exposures from several wavelengths in the same time duration are additive on a proportional basis of spectral effectiveness with due allowance for all correction factors. The simultaneous exposure to pulses and CW radiation is not linearly additive (there may be synergism) and caution should be used in these situations until more data are available.

The MPI values used in this standard are consistent with those of ANSI Z136.1. However, they have been modified for a 5 mm diameter pupil (ambient lighting) for wavelengths between 400 and 1400 nm. (For wavelengths between 1400 nm and 100 μm, the corresponding value of 1 mm is the same as that of ANSI Z136.1.) This value, which is the diameter of the limiting aperture, e.g., maximum circular area over which irradiance and radiant exposure is averaged, shall be used for measurements and calculations for all MPI values.

The MPI in terms of irradiance can be obtained by dividing the radiant exposure by time, t, in seconds; the MPI in terms of radiant exposure can be obtained by multiplying the irradiance by t.

9. Measurements

9.1 General. The procedures included in this standard are designed to minimize the need for irradiance measurements. It is generally sufficient to measure (when not known) the power in the fiber (closed-system measurement) since the AELs and exposure are related to this quantity. Normally, a potential hazard exists only when the end of an energized fiber is exposed during installation, service or during some splicing operations. In such cases, protective measures should be followed since the power level and wavelength of the OFCS transmitters or test sets may not be known. Furthermore, if power measurements are performed, the measurement process may be difficult since wavelength multiplexing may be used at widely different wavelengths. In addition to different ocular hazards associated with different wavelengths, the detector may respond differently to the different wavelengths.

9.2 Measurements of Power. If measurements are performed, the accuracy of the instrumentation should be traceable to an acceptable national standard, either directly to the National Bureau of Standards (NBS) or to other transfer standards traceable to the NBS. The NBS maintains a program for these purposes.

Measurements of the optical power emitted from the end of an optical fiber normally emphasize the collection of all optical power, not just that power within the angle defining the numerical aperture. A variety of specialized OFCS power meters have been developed and are commercially available. Optical power measurements for service classification for the purposes of this standard require the collection of total power (radiant flux) within the defining aperture specified.

10. Revision of American National Standards Referred to in This Document

When any of the following American National Standards referred to in this document is superseded by a revision approved by the American National Standards Institute, Inc., the revision shall apply:

American National Standard for the Safe Use of Lasers, Z136.1-1986

12

ANSI Z131.3, Laser Safety in Health Care Facilities

This standard, as stated in the Foreword, provides guidance for the safe use of lasers and laser systems for diagnostic and therapeutic uses in health care facilities. Once a laser or laser system is properly classified, there normally should be no need to carry out measurements or calculations to meet the provisions of the standard. ANSI Z136.3 was developed by the Committee as a part of the Z136 series because of the special environment in which lasers are used, including direct involvement with living tissue of patients. Only those sections deemed manditory for compliance are included here, with approval of the American National Standards Institute. The complete standard may be obtained from its publisher, the Laser Institute of America, 12424 Research Parkway, Suite 130, Orlando, FL 32826.

1. General

1.1 Scope. This standard provides guidance for safe use of lasers and laser systems for diagnostic and therapeutic uses in health care facilities. It is intended for use by all persons associated with the installation, operation, maintenance and service of Health Care Laser Systems (HCLS). This standard covers engineering, procedural, and administrative controls necessary for the safety of patients and health care professionals. These controls are based upon: evaluation of potential hazards from laser radiation; unique problems related to operating rooms, outpatient clinics, and private medical offices; and exposure risks to patients and professional staff. Included are engineering and administrative controls, personal protective equipment and laser safety training.

The control measures outlined herein shall not be considered to restrict or limit in any way the use of laser radiation, of any type, which may be intentionally administered to an individual for diagnostic, therapeutic, or medical research purposes, by or under the direction of qualified professionals engaged in the healing arts. However, those administering and assisting in the administering of the laser radiation, as well as the patient where applicable, shall be protected by control measures herein.

1.2 Application: Hazard Classification Scheme. The objective of this standard is to provide guidance for the safe use of diagnostic and therapeutic lasers for health care. The practical means for accomplishing this purpose is by classifying the HCLS according to its relative hazard and by specifying appropriate controls for each classification. The hazard classification scheme detailed in Section 3 of this standard is identical with that in ANSI Z136.1, and is based primarily on the ability of the beam to cause biological damage to the eye. Although hazards to skin and other parts of the body are of importance, they are associated only with Class 3 and Class 4 lasers and occur at exposure levels equal to or greater than those producing eye damage. Laser radiation as defined herein falls in the ultraviolet, visible, and infrared regions of the electromagnetic spectrum and should not be confused with ionizing radiation.

The laser hazard classification system is based entirely on laser radiation.

1.2.1 Laser Hazard Classification. It must be recognized that the classification scheme given in this standard relates specifically to HCLS and that the potential hazard of such instruments is based on their operating characteristics. However, the conditions under which the instrumentation is used, the level of safety training of the individuals who use and service the system, and other environmental and personnel factors are important considerations in determining the full extent of the safety control measures. Since such determination requires informed judgments by responsible persons, the major responsibility for such judgements in all locations where medical laser instrumentation is used shall be assigned to a qualified person, namely the Laser Safety Officer (LSO).

The recommended stepwise procedure for using this standard is as follows:

Determine the appropriate class of the laser system using the manufacturer's specification, or the hazard evaluation classification descriptions given in Section 3, or the methods described in ANSI Z136.1.

Note: HCLS certified for a specific class by a manufacturer in accordance with the Federal Laser Product Performance Standard may be considered as fulfilling all classification requirements of this standard.

Comply with the measures specified for the appropriate laser class using the following table as a guide. This procedure will, in most cases, eliminate the need for measurement of optical radiation, quantitative analysis of hazard potential, or use of the Maximum Permissible Exposure (MPE) values.

Class[a]	Control Measures	Medical Surveillance[b]
1	Not Applicable	Not Applicable
2	Applicable	Not Applicable
2a	Applicable	Not Applicable
3a	Applicable	Not Applicable
3b	Applicable	Applicable
4	Applicable	Applicable

[a]During normal operation and maintenance only. Alignment, maintenance and service procedures of an embedded Class 2, 3a, 3b or 4 laser shall require controls appropriate to the class during these functions. See Section 3.

[b]See Section 6.

For quantitative evaluation of the degree of hazard associated with a given laser system, the MPE values should be consulted. To use these values, first determine whether there is a possibility of intrabeam viewing (in which case the intrabeam MPE criteria in 8.2 shall be used), or whether a hazardous diffuse reflection is present (in which case the MPE values for an extended source or the intrabeam MPE values may be applicable). For the purpose of this standard, except for short distance viewing of small beam diameter or focused Class 3b lasers, only Class 4 lasers are considered capable of producing hazardous diffuse reflections, and hence calculations for viewing diffuse reflections from laser systems utilizing Class 1, 2, and 3 lasers are normally unnecessary.

HCLS can be divided into two broad hazard categories: those in the optical region, roughly 400 to 1400 nm, in which a focal image on the retina presents the main hazard; and those in the ultraviolet and infrared regions, in which the main hazard is to the cornea and skin. In general, Class 4 lasers and laser systems also present a skin and fire hazard and appropriate precautions should be taken. CO_2 laser surgery systems fall into this latter category. Almost by definition, surgical lasers and laser systems are Class 4, with a few exceptions at the higher levels of Class 3.

1.2.2 Federal Regulations. The Food and Drug Administration (FDA) Center for Devices and Radiological Health (CDRH) has the responsibility for implementing and enforcing the Federal Laser Product Performance Standard and the Medical Devices Amendment to the Food, Drug, and Cosmetic Act. These regulations have an important bearing on the regulatory status of HCLS.

1.3 Laser Safety Officer (LSO)

1.3.1 General. An individual shall be designated the LSO with the authority and responsibility to monitor and enforce the control of laser hazards, and to effect the knowledgeable evaluation and control of laser hazards.

An LSO should be designated for all circumstances of operation, maintenance, and service of Class 3a HCLS, and shall be designated for Class 3b or Class 4 HCLS.

Depending on the extent and number of laser installations, the position of LSO may or may not be a full time assignment. In some instances, designation of an LSO may not be required. Such instances could include operation and maintenance performed only on Class 1, Class 2 and Class 3a lasers. However, if service is performed on a laser product having an embedded Class 3b or Class 4 laser, the designation of an LSO may be required. In such instances the designation of an LSO may be the responsibility of the organization requiring access to the embedded laser or laser system, such as a service company or organization.

1.3.2 LSO Specific Responsibilities

1.3.2.1 Classification. The LSO shall classify, or verify classification of, lasers and laser systems used under the LSO's jurisdiction.

1.3.2.2 Hazard Evaluation. The LSO shall be responsible for hazard evaluation of laser treatment areas, including the determination of Nominal Hazard Zones.

1.3.2.3 Control Measures. The LSO shall be responsible for assuring that the prescribed control measures are in effect, recommending or approving substitute or alternate control measures when the primary ones are not feasible or practical, and periodically confirming the proper functioning of those control measures.

1.3.2.4 Procedure Approvals. The LSO shall approve standard operating procedures, alignment procedures, and other procedures that may be part of the requirements for administrative and procedural control measures. this shall include, for example, pre-, intra- and post-operative checklists used by operating personnel.

1.3.2.5 Protective Equipment. The LSO shall recommend or approve protective equipment, e.g., eyewear, clothing, barriers, screens, as may be required to assure personnel safety. The LSO shall assure that protective equipment is inspected periodically to ensure proper working order.

1.3.2.6 Signs and Labels. The LSO shall approve the wording on area signs and equipment labels.

1.3.2.7 Facility and Equipment. The LSO shall approve HCLS installation and equipment prior to use, and modification of existing facilities or equipment. The LSO shall authorize laser technicians for performance of maintenance and service.

1.3.2.8 Training. The LSO shall assure that adequate safety education and training is provided to HCLS area personnel.

1.3.2.9 Medical Surveillance. The LSO shall determine the personnel categories for medical surveillance.

3. Hazard Evaluation and Classification

3.1 General. Four aspects of HCLS influence the total hazard evaluation, and thereby influence selection of the control measures to be applied.

1. The capability of the radient energy of the HCLS to injure HCP or the patient's body area other than the intended treatment sites

2. The environment in which the HCLS is used

3. The HCP who may use, or be exposed to, laser radiation

4. The non-emission hazard associated with HCLS

Aspects (2), (3) and (4) vary with each laser application and cannot be readily standardized. The total hazard evaluation procedure shall consider all four aspects.

The HCLS classification scheme is based upon aspect (1). In this scheme, a Class 1 laser is the least hazardous class and is considered to be incapable of producing damaging levels of laser emission. Class 2 (low power) applies only to visible laser emissions and may be viewed directly for time periods less than or equal to 0.25 s, the aversion response time. Class 3 lasers may be hazardous under direct or specular reflection viewing. A Class 4 (high power) laser is hazardous to the eye from the direct beam and diffuse reflections. Class 4 lasers normally also present significant skin and fire hazards. Most surgical lasers, and any laser emitting in excess of 0.5 W average radiant power, are Class 4. See ANSI Z136.1, Section 1.2.

HCLS classified in accordance with this standard shall be labeled with the appropriate hazard classification (see 4.7). HCLS certified for a specific class by a manufacturer in accordance with the Federal Laser Product Performance Standard may be considered as fulfilling all classification requirements of this standard. It should be noted that in some cases there may be differences between this standard and the Federal Laser Product Performance Standard. If the laser has been modified for use in a manner other than originally intended that alters the basis for its classification. either reclassification or guidance by the LSO shall be used.

3.2 Laser Considerations. The wavelength of the laser must be known in order to confirm use of the appropriate control measures, i.e., whether retinal, corneal or skin hazards is the main safety consideration. In the event that multiple wavelengths or tunable lasers, e.g., dye lasers, are used, the LSO shall determine that appropriate safety measures are employed for all hazardous wavelengths to which the patient or HCP can be inadvertently exposed.

3.3 Laser and Laser Systems Hazard Classification. HCLS used in the health care environment will, in most cases, have their classification determined by the manufacturer. If the laser classification is not known, specific instructions for determining Laser Hazard Classification can be found in ANSI Z136.1, Section 3.3.

3.4 Operating or Treatment Rooms: Environment in Which the Laser is Used. Following HCLS classification, environmental factors require consideration. Their importance in the total hazard evaluation depends on the HCLS classification. The decision to employ additional hazard controls not specifically required in Section 4 is influenced by environmental considerations, principally for Class 3 and Class 4 HCLS.

4. Control Measures

4.1 General Considerations. In this section control measures are divided into administrative and equipment controls; laser treatment controlled area; maintenance, service and procedural controls; personnel protective equipment and warning signs and labels (see Table 1).

It is important to understand that the presence of the patient creates the need for a set of unique control measures which must be dealt with in special ways.

The ancillary hazards associated with the use of HCLS can be significant. For example, loss of life has occurred during electrical servicing and testing of laser equipment incorporating high-voltage power supplies. Fires, including explosions and endotracheal tube fires, have been reported.

In addition, airborne contaminants (gases, vapors and aerosols) and intense incandescent radiation (which can present additional eye hazards) may be released during the interaction of laser energy with materials, e.g., bone and carbonized tissue.

The control measures contained in this section shall apply when the equipment is in its intended operational mode. Control measures applicable to service are so noted. Engineering control measures which are incorporated into the HCLS itself or added to the installation by the user are almost always the preferred method for controlling access to laser radiation. If engineering controls are impractical or inadequate, administrative and procedural controls and personnel protective equipment shall be used. The limits of any type of control measure, e.g., failure modes of enclosures and eye protection, or the inability of some personnel to understand written warnings, shall be considered in developing a laser hazard control program.

Table 12.1 Control Measures for the HCLS

Classification	1	2a	2	3a	3b	4
Research (4.1.1)					X	X
Administrative controls (4.2)					X	X
Laser use committee (4.2.1)					X	X
Maintenance and service (4.2.2)					X	X
Standard operating procedures (4.2.3)					•	X
Authorized personnel (4.2.4)					X	X
Equipment controls (4.3)					X	X
Guarded switch (4.3.1)					X	X
Beam delivery disconnect safety features (4.3.2)					X	X
HCLS warning labels (4.3.3)		X	X	X	X	X
Service and repair of HCLS (4.3.4)	X	X	X	X	X	X
Modification of HCLS (4.3.5)	X	X	X	X	X	X
Quality control and safety audits (4.3.6)					X	X
HCLS output calibration (4.3.7)		•	•	•	X	X
Area posting sign (4.3.8)			•	•	X	X
Environment in which laser is used (4.4.1)	*	*	*	*	*	*
Laser treatment controlled area (4.4.2)					X	X
Laser treatment controlled area (4.4.2.1)					X	
Laser treatment controlled area (4.2.2.2)						X
Optical fiber surgical probes (4.4.3)	*	*	*	*	*	*
Alignment procedures (4.5.1)					•	X
Eye protection (4.5.2)					X	X
Service personnel (4.5.3)	X	X	X	X	X	X
Protective eyewear (4.6.2)					X	X
Protective clothing (4.6.3)	*	*	*	*	*	*
Warning signs and labels (4.7) signs / labels		/X	•/X	•/X	•/X	•/X
Explosion and fire hazards (7.5)						X

Legend: X - Shall
         • - Should
         * - LSO shall establish alternate controls

4.1.1 Research. In health-care research laboratories some engineering controls may not always be used. In such instances, the LSO shall establish alternate control measures which comply with the requirements set forth in ANSI Z136.1.

4.2 Administrative Controls. Experience has shown the necessity for a clear line of administrative control in safety matters. Thus, an LSO shall be appointed to effect administrative controls.

Administrative controls are methods and/or instructions which regulate work practices to implement or supplement engineering controls and which may specify criteria for personnel protection. The health care facility shall establish authorization and standard operating procedures (SOP). The procedures shall include, for example, pre- and post-operative checklists for use by operating personnel.

4.2.1 Laser Use Committee (Class 3b or Class 4). For institutions where a number of different practitioners may use Class 3b or Class 4 HCLS, a controlling body or laser use committee shall be formed to govern laser activity and establish use criteria.

4.2.2 Maintenance and Service (Class 3b or Class 4). The LSO has the responsibility to assure that maintenance and service for Class 3b or Class 4 HCLS shall be performed only by technicians trained in laser service, furnished either by the manufacturer's service organization or the institutional staff. In either case, the requirements and control measures of this standard and ANSI Z136.1 shall be followed during all aspects of service to HCLS.

4.2.3 Standard Operating Procedures (Class 3b or Class 4). The LSO should require approved written service and maintenance SOPs for Class 3b HCLS. The LSO shall require and approve written SOPs for Class 4 HCLS. These written SOPs shall be maintained with the HCLS for reference by the operator and maintenance or service personnel.

4.2.4 Authorized Personnel (Class 3b or Class 4). Class 3b or Class 4 HCLS shall be operated, maintained or serviced only by authorized personnel. HCLS with embedded Class 3b or Class 4 lasers shall be maintained or serviced only by authorized personnel, if such procedures would permit access to levels which exceed the appropriate MPE.

4.3 Equipment Controls. It is recommended that either engineering controls (as an integral part of the laser or laser system) or procedural controls be utilized where possible to avoid potential hazards associated with Class 3b or Class 4 HCLS. In addition to controls included on HCLS as required by the Federal Laser Product Performance Standard, the use of supplementary engineering controls should be considered. This may lower the class of the laser or laser system.

4.3.1 Guarded Switch (Class 3b or Class 4). For Class 3b or Class 4 HCLS, the switch which controls patient exposure shall be guarded, e.g., a guarded pedal switch or guarded finger-trigger switch. In environments with multiple-guarded foot pedal switches for different HCLS or other operating room instruments, precautions shall be taken to prevent inadvertent exposure.

4.3.2 Beam Delivery Disconnect Safety Features (Class 3b or Class 4). Laser beam delivery systems for Class 3b or Class 4 HCLS used in patient therapy, e.g., optical lightguides (articulated arms) or optical fibers, shall be considered as an extension of the protective housing. Disconnection of the delivery system from the laser source shall not permit laser radiation in excess of the applicable MPE to directly exit from the laser or laser system.

4.3.3 HCLS Warning Labels (all classes except Class 1). Labeling in accordance with ANSI Z136.1 or the Federal Laser Product Performance Standard shall satisfy the labeling requirements of this standard and any required labeling shall not be covered or removed from HCLS equipment.

4.3.4 Service and Repair of Laser Systems (all classes). The LSO shall ascertain whether any changes in control measures are required following any service or repair procedure which may affect the output power or operating characteristics of a laser or laser system so as to make it potentially more hazardous.

4.3.5 Modification of Laser Systems (all classes). The LSO shall determine whether any changes in control measures are required whenever deliberate modifications are made that could change the class of an HCLS and affect its output power or operating characteristics so as to make it potentially more hazardous.

4.3.6 Quality Control and Safety Audits (Class 3b or Class 4). A periodic laser safety audit of installation and equipment performance shall be conducted under the supervision of the LSO.

4.3.7 HCLS Output Calibration (Class 3b or Class 4). A confirming power/energy measurement shall be routinely performed using an appropriate radiometer as specified by the HCLS manufacturer, or as deemed appropriate by the LSO, to assure proper calibration of the HCLS power/energy monitor.

4.3.8 Area Posting Signs (Class 2, Class 3a, or Class 4). An area which contains a Class 2 or Class 3a HCLS should be posted with the appropriate sign as described in 4.7. An area which contains a Class 3b or Class 4 HCLS should be posted with the appropriate sign as described in 4.7. A notice sign, as described in 4.7, shall be posted outside a temporary laser controlled area.

4.4 Laser Treatment Controlled Area and NHZ (Class 3b or Class 4)

4.4.1 Laser Use Environment. The decision to employ additional control measures for Class 3 or Class 4 HCLS not specifically required in Section 4 of this standard, or to not implement some that are, is influenced by environmental considerations.

If exposure of unprotected personnel to the primary or specularly reflected beam is possible, determination of the irradiance or radiant exposure for the primary or specularly reflected beam at that specific location or of the radiance of an extended source is required.

When applicable, e.g., in the presence of unenclosed Class 3b or Class 4 beam paths, the NHZ shall be determined by the LSO. The NHZ describes the space within which the level of direct, reflected or scattered radiation, during normal operation, exceeds the applicable MPE. The NHZ may be determined by information supplied by the laser or laser system manufacturer, by measurement, or by using the appropriate NHZ equations or other equivalent assessment.

The LSO shall ensure that consideration is given to direct, reflected and scattered radiation in the establishment of boundaries for a laser controlled area. If operation or maintenance personnel are required within the NHZ, appropriate control measures shall be established. During service, a temporary laser controlled area may be required.

Since the levels of radiation that escape the NHZ are maintained at or below the appropriate MPE, no additional controls are required outside the NHZ.

Viewing the main beam or a specular laser target in the wavelength range of 0.3 to 4 μm with an optical instrument is potentially hazardous due to the instrument's light-gathering capability. Use of such optical systems without laser-safety filters may therefore effectively increase the NHZ boundaries and must be considered in the overall hazard analysis.

4.4.2 Laser Treatment Controlled Area (Class 3b or Class 4). A laser treatment controlled area shall be defined by the limits of the NHZ when the entire beam path of a Class 3b or Class 4 HCLS is not sufficiently enclosed and/or baffled to prevent access to radiation above the MPE. The laser treatment controlled area shall apply, but not be limited, to such applications as laser surgery, photodynamic therapy, and ophthalmic surgery.

4.4.2.1 Laser Treatment Controlled Area (Class 3b). A Class 3b laser treatment controlled area should contain the NHZ, the extent of which is clearly designated, and shall:

1. Be posted with the appropriate warning sign(s).
2. Be supervised and only be occupied by authorized personnel.

In addition to the above, a Class 3b laser treatment controlled area should:

3. Be under the direct supervision of an individual knowledgeable in laser safety.
4. Be so located that access to the area by spectators is limited and requires approval.
5. Have any potentially hazardous beam terminated in a beam stop of an appropriate material.
6. Have only diffuse reflective materials in or near the beam path, where feasible. The use of anodized or matte surfaces on instruments should be considered.
7. Have personnel who regularly require entry into a laser treatment controlled area adequately trained, provided with appropriate protective equipment and required to follow all applicable administrative and procedural controls.
8. Have all windows, doorways, open portals, etc., either covered or restricted in such a manner as to reduce the transmitted laser radiation to levels at or below the appropriate ocular MPE. Windows need not be covered for wavelengths where the glazing is opaque. Window glass is opaque from 0.18 to 0.3 μm and from 4 μm to 1 mm.
9. Require storage or disabling (removal of the key) of the laser or laser system when not in use to prevent unauthorized use.
10. Assure that the shutter mechanism is closed during stand-by periods.
11. Provide an emergency shut-off switch to enable rapid shutdown of equipment.

4.4.2.2 Laser Treatment Controlled Area (Class 4). A Class 4 laser treatment controlled area shall be designed to fulfill all requirements of 4.4.2.1 and, in addition, shall incorporate the following safety measures:

1. Entryway safety controls shall be employed for Class 4 laser treatment controlled areas. Three options are recommended below. The LSO shall make a choice based upon the following factors: training and knowledge of those routinely having access to the laser controlled area; extent of the NHZ; accessibility to the laser treatment controlled area by patients or the general public, e.g., possible access from facility hallways or patient waiting rooms; accessibility by untrained staff, e.g., in a medical research laboratory; availability of protective equipment, e.g., eyewear, prior to entry.

The LSO shall consider options (a) or (b), below, for areas, e.g., outpatient clinics, some medical research facilities, having a possibility of accidental unrestricted access by untrained persons.

The LSO shall consider options (b) or (c), below, for areas, e.g., surgery, limited access out-patient rooms, having a low probability of unrestricted access except by trained staff.

a. Non-Defeatable (non-override) Area Entryway Safety Controls. Non-defeatable safety latches, entryway or area interlocks, e.g., electrical switches, pressure sensitive floor mats, infrared or sonic detectors, shall be used to deactivate the laser or reduce the output to the appropriate MPE levels in the event of unexpected entry into the laser controlled area.

b. Defeatable Area/Entryway Safety Controls. Defeatable safety latches, entryway or area interlocks shall be used if non-defeatable area/entryway safety controls limit the intended use of the laser or laser system, e.g., during normal usage requiring operation without interruption such as during long-term testing, medical procedures or surgery. If it is clearly evident that there is no optical-radiation hazard at the point of entry, override of the safety controls shall be permitted to allow access to authorized personnel, provided that they have been adequately trained and provided with adequate personal protective equipment.

c. Procedural Area/Entryway Safety Controls. Where safety latches or interlocks are not feasible or are inappropriate, e.g., during medical procedures, surgery, the following shall apply:

    i. All authorized personnel shall be adequately trained and adequate personal protective equipment shall be provided upon entry.

    ii. A door, blocking barrier, screen, curtains, etc. shall be used to block, screen or attenuate the laser radiation at the entryway. The level of laser radiation at the exterior of these devices shall not exceed the applicable MPE, nor shall personnel experience any exposure above the MPE immediately upon entry.

    iii. At the entryway there shall be a visible or audible signal indicating that the laser is energized and operating at Class 4 levels. A lighted laser warning sign or flashing light are appropriate methods to accomplish this requirement.

2. All Class 4 area/entryway safety controls shall be designed to allow both rapid egress and admittance to the laser treatment controlled area under emergency conditions.

3. For emergency conditions there shall be a clearly marked "Panic Button" (remote controlled connector or equivalent device) available for deactivating the laser or reducing the output to the appropriate MPE levels.

4.4.3 Optical Fiber Surgical Probes. Back-scattering from frosted probes used with optical fiber delivery systems for some Class 3b or Class 4 HCLS can be hazardous, as is demonstrated by the extensive NHZ required at sites utilizing such probes. The LSO shall evaluate the application of such probes and implement personal and administrative controls commensurate with the potential hazard.

4.5 Maintenance, Service and Procedural Controls

4.5.1 Alignment Procedures (Class 2, Class 3a, Class 3b or Class 4). Experience has shown that a significant ocular hazard may exist during alignment procedures. Alignment of laser optical systems, e.g., mirrors, lenses, beam deflectors, shall be performed in such a manner that the primary beam, or a specular or diffuse reflection of a beam, does not expose an eye to a level above the applicable MPE.

Written procedures outlining alignment methods should be approved by the LSO for Class 3b and shall be approved by the LSO for Class 4 lasers or laser systems. The use of low power (Class 1 or Class 2) visible lasers is recommended for alignment of higher-power visible or invisible lasers and laser systems.

4.5.2 Eye Protection (Class 3b or Class 4).

4.5.2.1 General. Eye protection devices which are specifically designed for protection against radiation at the laser wavelength(s) in use shall be administratively required for all HCP within the NHZ for Class 3b or Class 4 HCLS and their use enforced when engineering or other procedural and administrative controls are inadequate to eliminate potential exposure in excess of the applicable MPE.

When long-term exposure to visible lasers is not intended, the applicable MPE used to establish the optical density requirement for eye protection may be based on a 0.25 s exposure.

When viewing an extended source or a diffuse reflection of a beam from a Class 4 laser or laser system where intermediate viewing time is intended, e.g., optical alignment procedures, the applicable MPE should be based upon the maximum viewing time anticipated during any given eight-hour period. For example, a maximum viewing time of 10 minutes is applicable for most alignment procedures when viewing a diffusely reflecting target.

If the extended source criterion is not applicable due to a small beam size, the exposure at the cornea from a diffuse reflector may be estimated using the inverse square law relationship and the intrabeam MPE criteria shall apply.

4.5.2.2 Patient Eye Protection. Appropriate eye shields should be placed on top of the cornea to protect the eye during laser treatment over the eyelid area. However, when facial areas, particularly around the eyes, are being treated, e.g., for port wine stains, protective glasses are not adequate to shield the eyes.

4.5.3 Service Personnel. Personnel who require access to the laser or laser system contained within a protective housing or beam enclosure for the purpose of service shall comply with the appropriate control measures of the embedded laser or laser system. The service personnel shall have education and training commensurate with the class of the embedded laser or laser system. See Section 5 for recommended training and education.

4.6 Personal Protective Equipment (Same as required in Z136.1)

4.7 Warning Signs and Labels (All Classes except Class 1).

4.7.1 Design of Signs. Sign dimensions, letter size and color, etc., shall be in accordance with American National Standard Specification for Accident Prevention Signs, ANSI Z35.1.

5. Laser Safety And Training Programs

5.1 Scope. Safety training programs shall be established for HCP using Class 3b or Class 4 HCLS and should be established for HCP using Class 2 or Class 3a HCLS.

Personnel that may require some form of laser and laser safety training include:

1. LSOs
2. Laser physicians
3. Laser technical support staff, e.g., clinical engineers, laser technicians
4. Medical support staff, e.g., nurses
5. Laser system service personnel, including either in-house or contractor service personnel

Each of these groups will require training responsive to the specific needs and responsibilities relating to the use of lasers.

5.2 Safety Program. The health care facility or private practitioner, as applicable, shall establish and maintain an adequate program for the control of laser hazards. This program should include the following:

1. Delegation of authority and responsibility for the supervision of evaluation and control of laser hazards to an LSO. Where the number or diversity of laser usage warrants, an associated safety or laser use committee shall be formed to govern laser activity and establish use criteria.
2. Authorization and control of HCP as well as other personnel who enter the NHZ shall be effected by the safety administration in order to ensure adequate knowledge of safe procedures and control measures appropriate to each individual function. Many health care facilities find it convenient to incorporate HCLS user authority in credentialing procedures connected with granting privileges.
3. Application of adequate protective measures for the control of laser hazards as required in Section 4, Control Measures.
4. Management of accidents, including reporting accidents and preparing action plans for the future prevention of accidents following a known or suspected incident. This may be accomplished through the use of internal incident control reports.

5. Education of authorized personnel (operators and others) in the assessment and control of laser hazards. This may be accomplished through training programs.

5.3 Education. Detailed training in laser safety procedures shall be provided for those HCP, e.g., surgeons, physicians, using Class 3b or Class 4 HCLS or those responsible for patient care, e.g., anesthesia personnel, nursing staff, during the procedure. Training shall also be provided for those on the technical support staff responsible for pre- and post-surgical laser system checkout as well as system monitoring during surgery, e.g., laser technicians, nurses, administrators, clinical engineers, medical physicists. The training shall provide a complete understanding of the requirements of a safe laser environment.

6. Medical Surveillance of Health Care Personnel (HCP)

6.1 General. Medical surveillance is not required for HCP using Class 1, Class 2, Class 2a or Class 3a HCLS. The type of medical surveillance that shall apply depends on the personnel category. The rationale for medical surveillance requirements for HCP is threefold.

1. To ensure a baseline level of visual performance against which damage can be assessed in the event of a laser accident, thus formulating possible prognosis for recovery. This is particularly oriented towards ocular hazards.
2. To identify those individuals who may be at special risk from ultraviolet hazards. This portion of medical surveillance is primarily for dermatological hazards.
3. To supply a baseline level of ocular performance at initiation and termination of exposure and environment, i.e., pre- and post-employment physical examinations.

Medical surveillance requirements have been limited to those HCP who have possible exposure to hazardous laser radiation from HCP. Any medical surveillance program should be individualized so that HCP examinations are appropriate to the known risks of particular types of laser radiation in use at the various locations throughout the health care facility. Some HCP may need an additional examination for medical/legal reasons, or to conform with established principles of what constitutes a comprehensive ocular or dermatologic examination, or as part of planned epidemiological studies.

Further details of the rationale and specific information of value to examining or attending physicians are included in ANSI Z136.1, Appendix E.

10. Revision of American National Standards Referred to in This Document

When any of the following American National Standards referred to in this document is superseded by a revision approved by the American National Standards Institute, Inc., the revision shall apply:

American National Standard National Electrical Code, ANSI/NFPA 70-1987

National American Standard Fundamentals Governing the Design and Operation of Local Exhaust Systems, ANSI Z9.2-1979

American National Standard Practices for Respiratory Protection, ANSI Z88.2-1980

American National Standard Safety Standard for Radio Receivers, Audio Systems, and Accessories, ANSI/UL 1270-1978

American National Standard for the Safe Use of Lasers, ANSI Z136.1-1986

American National Standard Specifications for Accident Prevention Signs, ANSI Z35.1-1972

American National Stnadard Practice for Occupational and Educational Eye and Face Protection, ANSI Z87.1-1979

Hazards Associated
with Lasers

In addition to radiation hazards inherent in laser beams, other forms of potential danger lurk in various components of lasers systems. Electrical energy is utilized in essentially all lasers. Chemical hazards abound. Cryogenics, bottled gas, combustibles, x-rays, and especially mechanical hazards are all associated with laser operations. Much of the following informative hazard control data has been extracted from a document prepared by the author for the Physics Division of the Los Alamos National Laboratory and issued in 1981. In addition to describing control of laser-associated hazards, safe laboratory operating procedures are discussed.

ELECTRICAL SAFETY

The rules governing activity around electrical equipment must be reviewed regularly to prevent possible accidents. Employees assigned to modify or maintain electrical gear should be familiar with established rules and policies as follows:

1. Assume that all electrical circuits are energized until positive action is taken to ensure lack of power.

2. Do not handle electrical equipment when hands, feet, or body are wet or perspiring, or when standing on a wet floor.

3. Regard all floors as conductive and grounded unless covered with well-maintained and dry rubber matting of suitable type for electrical work.

4. Whenever possible, use only one hand when working on circuits or control devices.

5. When you must touch electrical equipment (for example, when checking for overheated motors), use the back of the hand. Thus, if accidental shock were to cause muscular contraction, you would not "freeze" to the conductor.

6. Wear safety glasses where sparks or arcing may occur, such as opening or closing high voltage switches.

7. Avoid wearing rings, metallic watchbands, etc. when working with electrical equipment or in the vicinity of strongly induced fields.

8. Provide overhead runways for extension cords and other plug-in receptacles to keep all electrical leads above floor level and out of walkways.

9. Label principal disconnectors.

10. Learn the following *rescue procedures* for helping a victim of electrical shock.

    a. *Kill the circuit* (know location of power switches).

    b. *Remove victim* with a nonconductor if still in contact with an energized circuit or if status of circuit is unknown. Use a laboratory coat as a loop, or your trousers, but *hurry!*

    c. *Administer first aid*, including cardiopulmonary resuscitation (CPR) if necessary. Training in life-safety courses is recommended for anyone working with electrical circuits.

    d. Have someone *call for emergency medical aid.*

Power Distribution Lines

Most companies have a policy that requires that all power lines be installed, maintained, and replaced by journeymen electricians. However, employees may perform certain circuit-breaker functions in laboratories that contain high-voltage systems. The following advice is offered for use in operating circuit breakers.

1. Do not touch anything except operating handles.

2. Use only one hand.

3. Wear protective eye gear.

4. Turn face away from breaker.

5. Push (or pull) with a fast motion.

6. Do not close breaker until equipment is ready for energizing.

7. Notify all persons concerned that the circuit will be activated.

Capacitors

Capacitors or capacitor banks having an impulse capability of 0.25 J or more constitute an *electrical shock hazard*. Those having *50 J* or more carry a potential *lethal shock hazard*.

Types of hazards

1. Even after having been disconnected and discharged, capacitors may again build up an electrical charge because of the phenomenon of "dielectric absorption." Passing atmospheric electrical disturbances can leave an electrical charge on a capacitor.

2. Shorting discharge can cause electrical arcs and the resultant energy release can cause burning by radiation, heat, or flying molten material.

3. Internal faults within capacitors often result in ruptured containers.

4. If an internal fault occurs, causing rupture of the case, or if leakage is ignited by an electrical arc or other source of ignition, a liquid dielectric capacitor can create a serious fire hazard.

5. Fuses are frequently used to protect individual capacitors from the total energy of the capacitor bank. If not adequate, such fuses and capacitors could explode, throwing dangerous projectiles.

6. In relation to all above hazards, capacitors used in inductive circuits can, during transient conditions, develop a charge or overvoltage.

Safety considerations for capacitors

Because capacitors offer so many potential hazards, a nearly complete checklist that should be considered in any effort involving capacitors, particularly in activities that increase energy levels, has been compiled:

1. Provide fault-current-limiting devices, such as fuses or resistors, capable of clearing or dissipating the total energy.

2. Provide protection against projectiles that may be produced during faults by the use of suitable enclosures and barriers.

3. Provide enclosures designed to prevent accidental contact with terminals, cables, or exposed electrical contacts.

4. Provide a locked or interlocked grounded metal enclosure.

5. Prevent or contain fires by reducing combustible material in the vicinity of the capacitors.

6. Automatically dump (crowbar action) capacitors before opening any access door.

7. Provide an appropriate (short) discharge time-constant in the grounding system device.

8. Check that each capacitor is discharged, shorted, and grounded prior to allowing general access to a capacitor area.

9. Provide reliable grounding, shorting, and interlocking devices.

10. Install "crowbars," grounding switches, cables, and other safety devices to withstand the mechanical forces that can occur when faults or crowbar currents flow.

11. Provide suitable warning devices, such as signs and lights.

12. Place shorting straps on each capacitor during maintenance periods while capacitors are in storage.

13. Provide manual grounding equipment that has the connecting cable visible for its entire length.

14. Supply safety devices, such as safety glasses, rubber gloves, and insulating poles.

15. Provide metering, control, and auxiliary circuits that are suitably protected from possible high potentials, even during fault conditions.

16. Perform routine inspection for deformed or leaky capacitor containers.

17. Provide a grounding stick that has a discharge resistor at its contact point, an insulated ground cable (transparent insulation preferred), and the grounding cable permanently attached to ground. Such a grounding stick should not be used to ground an entire large-capacity capacitor bank. Large-capacity shorting bars (with resistors) should be part of the stationary equipment.

Electronic Equipment (Including Power Supplies)

The instrumentation discussed in this section includes circuits and equipment used in measuring, monitoring, observing, and recording mechanical, electrical, and chemical phenomena. The instrumentation and controls to be considered are limited to those used with systems that are either operated at more than 300 V above ground or where stored energy exceeds 15 J.

The power supplies being considered are limited to those having either ac or dc outputs exceeding 600 V terminal-to-terminal, or the capability of operating at a potential in excess of 300 V above ground.

Instrumentation and controls

The following types of hazards can occur with the electronic gear described above.

1. If controls or interlocks fail, personnel can gain uncontrolled access to hazardous areas. These failures can result either when interlocks and associated relays are damaged electrically or mechanically, or when such interlocks and relays are bypassed by operating personnel.

2. False signals or erroneous instrumentation readings can result in hazardous conditions.

3. Excessive voltage on instrumentation or controls due to induced voltages or contact with high-voltage components can be hazardous to operating personnel.

4. Improper coordination of protective fuses or circuit breakers can result in overheating of electrical conductors because of electrical faults or overload.

5. Hazards can be created by the introduction of instrumentation equipment into a system that does not necessarily require the instrumentation for its operation.

   Examples

   a. A high-intensity stroboscopic light used for photography may adversely affect eyes, ignite flammable materials, and may require high voltages.

   b. Voltage-dividing networks may, under normal fault conditions, cause the investigator's measurement device to operate at high-voltage levels between input terminals to ground, or to other equipment.

   c. "Floating" chassis and/or ground loop circuits can result in potentials or current levels far in excess of those anticipated.

   d. Under normal and/or fault conditions, the instrumentation system may introduce, or be the subject of, energy levels in excess of the ratings of either or both circuits or of personnel safeguards.

6. Protective relay coils and their electrical contacts, if not sufficiently rated for the circuit, can burn open or weld closed on a momentary short circuit and thus leave the protective circuit inoperative for future operations.

Safety precautions to control these hazards are:

1. Provide the electrical control and instrumentation circuitry with adequate isolation at its interface with the main power equipment being controlled and monitored, considering both the normal and the fault conditions that may exist during operation of the main equipment.

2. Provide isolation where any hazzardous situation exists, by using isolation devices such as transformers, high-impedance components, or telemetering equipment such as radio, light, or sound.

3. Provide relays and interlocks on instrumentation and protective circuits with contacts rated at least as high as the voltage of the circuit and with current ratings as high as the protective fuse or circuit breaker used in the circuit. Observe carefully the inductance of the circuit in the proper application of relays and interlocks.

4. Design the circuit in a failsafe manner so that the loss of power does not result in a hazardous condition.

5. Provide a systematic procedure for *only* authorized personnel to obtain permission to bypass interlocks when necessary.

6. Provide systematic operating procedures to inform affected personnel of interlock bypass conditions. The system must not remain in the bypassed condition longer than necessary.

7. Set up procedures that allow only qualified personnel to make connections to high-voltage systems.

8. Provide redundant control, indication, or instrumentation on sections of a system where an operating error or faulty instrumentation could otherwise result in a hazard.

9. Provide a clear indication of the status of remote controlled equipment, with positive feedback for each specific command.

10. Provide shorting devices for use on current transformers when connecting or disconnecting instruments or controls.

11. Provide sufficient isolation devices and barriers between high-voltage and low-voltage equipment.

12. Provide proper overload protection on control circuits; and prevent low-current rated conductors from contacting circuits that have only large overcurrent protection.

13. Wire control circuits to prevent "sneak" circuits and accidental grounding of one line from causing safety devices to become inoperative.

14. Provide a consistent labeling procedure for control buttons, knobs, etc., and encourage the use of mimic control diagrams.

15. Route control wires so that no large looped (ground or otherwise) circuits are formed.
16. Arrange controls for large systems so that unit control circuits are electrically isolated from, but subordinate to, an overall system control.
17. Test controls by simulating failures or maximum-limit features before placing system into operation.
18. Provide visible indicators showing the bypassed interlocks, which should set automatically after one cycle.
19. Schedule regular inspections of interlocks and document conditions.

Power supplies

The types of hazards likely to be created by the use of power supplies in research facilities require the following precautions:

1. Make input connections to power circuits through either manual or automatic switching devices. Consider providing overcurrent, undervoltage, or other protection, depending on type of load supplied.
2. Provide proper isolation devices or physical barriers to prevent high-voltage stored energy from being dissipated in the low-voltage supply and/or control circuits. Consider two means of isolation so that the failure of one does not result in injury or excessive damage.
3. Install all high-voltage components in separate, isolated enclosures. In addition, interlocking may be required to render the high-voltage areas inaccessible unless the input power is de-energized, and access should be limited to qualified personnel only.
4. Provide automatic switches or contacts in output circuits, as well as manual devices, for grounding and shorting the power supply when the power has been turned off and when personnel are allowed access to an area containing high voltage.
5. The power supply system should contain a minimum amount of flammable liquid. Construction features should provide for self-containment of the liquid.
6. Make input and output connections at terminals that are covered or inaccessible during normal operations.
7. The main input supply switch should be located within sight of the power supply, and clearly identified.
8. Use alarms, such as signs or lights, to warn personnel and to indicate an energized power supply, especially on remote loads or power supplies.

9. Limit the number of startup stations or controls and provide shutdown stations or emergency stop controls at all remote locations.

10. Construct the enclosure or protective barriers of noncombustible conducting material and ground it to the adjacent building steel with a conductor suitable for fault conditions.

11. Use overload and short-circuit protection for the power supply output. These should operate the disconnect switches in the input circuit, as per item 1, and output grounding and shorting facilities, as per item 4.

12. To prevent overheating and fires, remote control and monitoring circuits should have overload or short-circuit protection.

13. Remote control and monitoring circuits subject to high voltage or high energy should be suitably isolated, or a provision should be made to prevent personnel injury or equipment damage.

14. For inductive loads, use "freewheeling" diodes or thyrite elements connected across the power supply dc terminals to ensure satisfactory discharge of stored energy.

15. Prior to initial operation and as a maintenance procedure, "megger" or "pot" the power supply; calibrate and check all protective devices.

NOTE: For additional electrical safety information, see *Practical Electrical Safety* by D. C. Winburn, published in 1988 by Marcel Dekker.

CHEMICAL SAFETY

Any chemical can be stored, handled, and used in a safe manner if its hazardous physical and chemical properties are understood and if the proper safeguards and personal protective equipment are used.

Storage of Chemicals

In general, store heavy items on or as near the floor as possible; apparatus and glass tubing should not project beyond front shelf limits. Provide stops or clamps for bulky or fragile apparatus to prevent persons accidentally jarring or brushing them.

To reduce the danger of chemical spillage, install wooden or metal ridges on the front of a storage shelf.

Plan special care in grouping liquid reagents to circumvent the danger of hazardous combinations of heat. Chemicals that might react together to produce dangerous fumes, fire, or explosion should be stored remotely from each other. Keep volatile liquids away from heat sources and sunlight.

Many laboratories now store all glass bottles of flammable liquids, hydrogen peroxide, acids, caustics, and similar materials within metal or stoneware receptacles sufficiently large to contain the entire volume in event of glass breakage.

Cap cylinders of gases in storage, support them to prevent rolling or toppling over, and place them away from heat and open flames.

Do not permit even the smallest amount of waste to accumulate on the floor surface or in corners. Advise janitors and maintenance men of the precautions to observe in stock and storage areas.

All containers, including compressed gas cylinders, should be labeled as to contents before they leave the chemical warehouse.

Handling of Chemicals

For easier handling of carboys and other large vessels, use bottle tilters. Rubber gloves, aprons, and safety goggles or face shields should be available from stock; and the use of these protective items is demanded of everyone working with hazardous chemicals.

The use of mechanical devices, such as lifts, trucks, and carts eliminates the manual lifting of unwieldy crates, bottles, and cans. If a drum pump is not used, drums should be mounted horizontally, securely braced to prevent rolling, and equipped with small, sand-filled boxes placed on the floor under the taps to absorb drippings. A faucet with spring closing action and locking pin is recommended.

Transport acids and flammable liquids, packaged in glass containers, in suitable safe carrying cases which should be available from the stockroom.

Flammable liquids that are dispensed from large containers should be carried in safety cans. Place the label on the container before it is filled; never fill it with material other than that called for by the label.

If possible hazard information is not received with a shipment of new or unfamiliar types of chemicals for special work, request the safety office or the industrial hygiene group to furnish comments and recommendations before proceeding with laboratory use.

In transporting heavy cylinders of compressed gases, use a hand-cart equipped with chains or straps. Never move two cylinders at the same time unless they are properly banded together or placed on a double cart. Do not bounce or jostle cylinders. Immediately after positioning them for use, chain or clamp the cylinders to prevent shifting or toppling. Special support rings for multiple cylinder support should be available from stock.

The following precautions are important in *handling compressed gas cylinders.*

1. Do not subject any part of a cylinder containing a compressed gas to a temperature above 125°F. Do not use flames or torches to heat cylinders. Store tanks in shade.

2. Do not tamper with the safety relief devices in valves or cylinders, and do not attempt to repair or alter cylinders.

3. Do not use cylinders as rollers, as supports, or for any purpose other than to carry gas.

4. Keep cylinder valves closed at all times except when actually using gas.

5. Where caps are provided for valve protection, keep the caps on cylinders when moving them.

6. Where cylinders are connected to a manifold, the manifold should be of proper design and equipped with appropriate regulators and relief valves.

7. Do not use regulators, gauges, hoses, and other appliances provided for use with a particular gas or group of gases on cylinders containing gases with different properties.

8. Open cylinder valves slowly, pointing the valve opening away from you. If a cylinder is difficult to open, do not use force. Return the cylinder to stock with an appropriate tag.

9. Do not use compressed gas to dust off clothing, as this may cause serious injury to the eyes or body.

10. Do not use compressed gases where the cylinder is apt to be contaminated by the feedback of process materials unless protected by suitable traps or check valves.

11. Before a regulator is removed from a cylinder, close the cylinder valve and release all pressure from the regulator.

12. Keep ignition sources away from cylinders of flammable gases.

13. Segregate reserve stocks of cylinders containing flammable gases from oxygen cylinders.

14. Be knowledgeable about properties of gas to be used.

Use care when handling materials such as boxes, crates, drums, or cylinders.

1. Wear serviceable gloves and shoes.

2. Know the safe way to lift. Keep your back straight and lift with your leg muscles.

3. Do not lift any package without some idea as to how heavy it is. In general, do not attempt to lift more than 50 pounds by yourself.

4. Look out for slivers, nails, wire, and projecting metal straps on packages being handled.

5. Get a good grip on the package and check your footing.

6. Do not use cardboard or fiber boxes in damp or wet places.

7. Glass bottles are easily broken and should be kept separated from each other.

Proper Techniques for Handling Chemicals

To reduce the chance of accidents, use correct techniques when assembling apparatus or when handling materials.

Flush the outside of acid bottles before opening them. Do not lay the cap down on any surface where a person may rest his hand or arm; and make certain that no spillage remains on tables, floor, or bottle.

Certain chemicals are packaged in special protective packing because of their hazardous nature; for example, hydrofluoric acid, hydroiodic acid, and the fuming acids. Opening such specially packaged chemicals involves special techniques, which may involve prior cooling of the bottle. For the proper procedure to use, contact your supervisor or the safety office.

Mouth pipetting is prohibited. Many bottles containing liquid chemicals, on being opened, act as though they were pressurized. Wear rubber gloves when opening bottles containing acids or caustics, and cover the dap with a piece of cheesecloth or paper towel until the cap is unscrewed.

Chemical and Other Burns

Incidental burns, along with minor cuts and scratches, are the most common of all injuries resulting from laboratory accidents.

Spatterings from acids, caustics, and strong oxidizing agents on skin or clothing should be flushed off immediately with enough water. Eye-wash devices and emergency showers are required in laboratories where active chemicals are used.

There are many safeguards for chemical or heat burns: approved mittens, tongs, rubber gloves, chemical hand creams, aprons, face shields, goggles, and emergency showers.

Wear gloves or a mitten, or use a pair of tongs when handling hot objects. Safety measures become second nature to the careful worker.

When heating liquids on a hot plate, use beaker covers to prevent spattering. Keep a pair of tongs conveniently at hand.

Keep bench tops wiped clean. One drop of acid may cause severe burns. Know the materials you are handling. Anticipate results—do not proceed without caution and forethought.

Burns can also be caused by ultraviolet (UV) light, infrared (IR) rays, and extremely cold materials such as dry ice and cryogenic fluids. Use proper goggles when working with UV or IR light and be careful to limit your time of exposure.

Use special techniques when opening bottles of hydrofluoric acid (HF). Place bottles in a laboratory refrigerator or under a stream of cold water for at least 15 minutes before opening in an approved chemical hood. Wear rubber gloves and eye protection.

Epoxy Resins

The materials contained in the various epoxy resin formulations could be quite hazardous. The amine curing agent or catalyst is a primary skin irritant and is a sensitizer to many persons. Inhalation of the vapors of curing agents can produce severe inflammation of the throat.

The following information is presented as a general summary of the recommended procedures and precautionary measures to be observed in the use of epoxy resins and their formulations. Specific detailed recommendations must be considered for each individual operation, and where questions arise they should be referred to the safety office.

Some suggested hygiene practices are:

Personal hygiene. The most important measure in combating dermatitis is the practice of *good personal hygiene*, and can be accomplished by:

1. Critical evaluation of work procedures to minimize skin contact with solvents, curing agents, and formulated mixtures
2. Provision of wash bowls, mild soaps, clean towels, and refatting creams

Protective devices. To minimize personal contact, wear:

1. Rubber gloves and protective sleeves (discard contaminated gloves)
2. Protective clothing such as coveralls (for all workers who have repeated or prolonged contact), rubber aprons as needed for additional protection, and approved footwear (do not wear contaminated clothing)
3. Full face shields, to ensure eye protection, whenever the possibility of airborne contamination exists

Handling procedures. Establish procedures for mixing and application to minimize skin contact and vapor exposure, as follows.

1. Mix the epoxy resin with the curing agent or solvent in a properly ventilated hood.
2. Confine curing operations to area provided with exhaust ventilation facilities.
3. Practice good housekeeping. Clean work areas daily; cover work benches and floors with wrapping paper or other disposable material. The materials should be mixed in cardboard or other disposable containers using refuse cans with foot-controlled lids. This will avoid the necessity of frequent general cleanups with solvents. Cases of dermatitis have been reported from accidental exposure by contact with spillage or contaminated tools.

The importance of observing the above recommendations increases proportionately with elevated temperatures of both the formulated materials and the atmosphere of the work stations as well as the work load.

Toxic Dust, Fumes, Gases, and Volatile, Flammable Solvents

Exposure to harmful air contaminants is a hazard to the health of laboratory workers, although it is one of the least generally recognized.

A good whiff of poisonous gas may cause serious illness; yet people neglect to take precautions or to consider the cumulative effect of daily exposure to toxic air contaminants at relatively low concentrations.

In every procedure that may produce toxic or poisonous air contaminants, apply local exhaust ventilation to remove harmful materials at their source.

Approved exhaust ventilation equipment includes chemical fume hoods, dry boxes, slot-type bench hoods, and so on, but does not include canopy hoods. A canopy hood is not recommended except for removal of heat, water vapor, or other nontoxic materials.

Do not use laboratory fume hoods for general storage. The minimum face velocity of air at the front of a hood for nonradioactive materials should be 100 ft/min. Make periodic checks of all fume hoods to determine their effectiveness; and place a label on the hood, indicating the hood is "satisfactory" (effectively exhausts toxic materials, etc.) or "unsatisfactory."

The fundamental rule to remember when working with volatile, flammable materials is that heat causes expansion and that confinement of expansion can result in explosion. Many chemical reactions are exothermic and produce spontaneous hazards so that danger exists even though external heat may not be available.

When evaporating volatile, flammable solvents in drying ovens, check the temperature regulators of the ovens often. Ensure that excessive pressure in ovens and furnaces is vented.

When working in a laboratory where poisonous chemicals are used, assumptions can result in severe consequences; for example, assuming the contents unlabeled containers are not caustic or flammable.

Do not drink from a beaker. Never taste any chemical. Smell a chemical only when necessary and then only by directing a small amount of vapor with the hand toward the nose.

Keep *mercury* confined in secondary, unbreakable containers. Report any spillage immediately to an appropriate health authority.

Industrial hygienists are responsible for assisting in the protection of personnel against toxic dusts, fumes, and gases. These specialists can evaluate potential exposures by air sampling and analysis; and, because they have the latest toxicity data, they should be consulted for advice on the toxicity of new or unknown chemicals.

Beryllium

The occupational exposure limits for beryllium is $2 \ \mu g/m^3$ of air, which is approximately the same linit for airborne dust or fumes.

Do not start any work with beryllium or its compounds without the approval of industrial hygiene regulations.

Solvents

Solvents may be toxic by inhalation and skin contact as well as flammable. Carbon tetrachloride therefore is not a "safe" solvent. When one can detect its odor, the air concentration is about three times the permissible level for an 8-hour day. Carbon tetrachloride, benzene, and carbon disulfide are the most toxic solvents in common use. Where possible, substitutes should always be used, especially for cleaning. It is possible to substitute ethanol or freon TF, or methyl chloroform (vythene) for carbon tetrachloride. It may be possible to substitute toluene or xylene for benzene. Use of benzene in volumes above milliliter levels may require approval. When using large quantities of any solvent, contact the safety office for air sampling to ensure contamination levels within current criteria.

Silvering Solutions

In all commonly used methods of chemical deposition of silver on glass, the silver is in an ammoniacal solution prepared by adding ammonium hydroxide to a solution of silver nitrate, and the silver is subsequently precipitated in the metallic form by the addition of a suitable reducing agent. In this solution explosive compounds may be formed, which detonate as a result of the slightest mechanical disturbance. such explosions have resulted in the loss of sight and have been sufficiently violent to cause considerable property damage. Explosions are possible with any of the methods of silvering but can occur when using a formula in which potassium hydroxide is a constituent. Explosions can occur if the residues remaining after silvering are allowed to become dry, or if empty vessels that have contained silvering solutions are stored without first being carefully washed. Consequently, it is advisable to dispose of all silvering solutions as soon as possible after the work is finished, and wash all vessels. If silvering is done on such a large scale that it is desirable to recover the silver from the solution, hydrochloric acid should be added to acidify the solution and precipitate the silver. Wear goggles during silvering work.

Volatile, flammable liquids

Do not discard volatile, flammable liquids by pouring them into a sink. An oxidizing agent may be put into the sink soon afterward, or the vapors may flow through the drain line and escape from the trap beneath another sink, causing them to ignite at a distant point. Contact the safety office for proper disposal procedures.

Careless mixing of chemicals can cause accidental fires. Flammable liquids such as ethyl ether and acetone are easily ignited by nitric acid or fuming sulfuric acid. The mixing of flammable liquids with liquid oxygen or liquid air can cause fires or explosions. Hydrogen sulfide has been reported to ignite in contact with a mere trace of sodium peroxide. Do not put volatile flammable liquids into drying ovens, even though the general oven temperature is low. Most laboratory ovens are heated with bare resistance elements operating at a temperature higher than the ignition temperature of the flammables.

The following compounds are flammable, with a flash point of 80° F or less and require a red label for shipping purposes: acetaldehyde, acetone, benzene, butanealdehyde, carbon disulfide, crotonaldehyde, dimethyl cellulose, dimethyl furan, dioxane, ethanol, ethyl acetate, ethyl buteraldehyde, ethylene dichloride, ethylene oxide, ethyl ether, ethyl propionate, heptane, hexane, isopropanol, isopropyl acetate, isopropyl chloride, isopropyl ether. methyl pentane, methyl propyl ether, propylene dichloride, propylene oxide, toluene, vinyl acetate, and vinyl chloride.

Regard open gas flames as an instant source of ignition for accidental fires. When adjusted to give maximum heat, they are essentially nonluminous and easily overlooked. Extinguish all flames not in use; turn the gas off at the bench or wall valve, not at the burner itself. Electric hot plates do not appear to have the intensity required to ignite flammables, and due to this misconception, accidental laboratory fires have been started. When working around flammables, treat electric hot plates just like burners, and consider timer controls.

Flammables can be ignited by electric sparks from contacts in switches, motors, and other electric equipment as well as from static charges. Electric equipment used in the presence of flammables should be vaporproof or explosionproof.

If an accident occurs in which one's clothing is set afire, avoid running, since running fans the flame. Wrap in a blanket—wrap the blanket around the neck first; drop to the floor, and roll over slowly. If no blanket is available, drop to the floor and roll over slowly. Avoid inhaling any of the flame, put the left hand on the right shoulder and the right hand on the left shoulder, and pull the arms against the face for protection. Get under a shower or otherwise douse the body and clothing with water, or roll in spilled water on the floor.

Oxygen and oil, and oxygen and other combustible materials, are spontaneously explosive. For this reason, equipment in which oxygen is handled should not be lubricated with hydrocarbons or other flammable materials. Jets of oxygen should not be permitted to strike oily or greasy clothes, or to enter a tank or vessel that has contained any flammable substance.

Chlorine and ethylene have been observed to explode spontaneously when mixed. Certain fluorine mixtures are also explosive hazards. Perchloric acid has a strong tendency to explode when evaporated in contact with organic or combustible matter. Magnesium perchlorate, which is often used for drying inert gases, is known to have caused explosions when contacted by combustible gases. Ethyl ether peroxide is formed by the oxidation of ethyl ether that is not inhibited. Storing the ether under an inert atmosphere (for example, nitrogen) or in contact with copper prevents formation of this peroxide. Ethyl ether used in the laboratory can be stabilized against peroxide formation by keeping a fair-sized piece of copper wire in the container. Peroxide concentrates in residues on distillation, and an explosion is likely to result if the evaporation is continued to dryness. Isopropyl ether has properties similar to ethyl ether and has a strong tendency to form isopropyl ether peroxide, but this is not retarded by copper. Certain organic antioxidants such as mono-benzyl paramino phenol prevent its formation.

Acetylene gas in cylinders is dissolved in acetone and absorbed in solid material having a large surface area. Acetylene gas compressed alone is explosive, because it tends to polymerize under pressure, and the reaction is exothermic. Therefore, acetylene should not be compressed except under carefully defined and controlled conditions. In general, acetylene gas should not be used at pressures in excess of 15 psig. Under certain conditions, acetylene forms explosive metallic compounds of copper, silver, and mercury, called acetylides. Dry copper acetylide is readily detonated by heat or shock. Copper acetylide is soluble in and quietly decomposed by HCl, hence copper equipment that has been in contact with acetylene or other equipment in which acetylides are suspected should be rinsed with HCl before welding or heating.

Hazardous Chemicals

Strong oxidizing agents such as nitric acid and nitrates, permanganates, peroxides, perchlorates, and perchloric acid must be handled with great care to avoid mixture with easily oxidizable materials such as organic matter, sulfides, etc. When dry, such mixtures may be explosive.

Chlorates and perchlorates are dangerous because of the oxygen liberated from these compounds when heated and also because they form explosive mixtures with sugar, charcoal, shellacs, starch, sawdust sweepings, and other organic matter. Explosive mixtures are also formed with sulfuric acid, potassium cyanide, phosphorous, and antimony.

Keep in mind the potentially destructive power of perchloric acid. Personnel using perchloric acid should wear protective rubber aprons, gloves, eye shields, antisplashing goggles, etc. Dispose of discolored acid immediately, that is, acid that has been contaminated by an outside source.

Do not use or store perchloric acid in or around wooden benches or tables, and use it only in approved posted hoods. Keep the bottles on a glass or ceramic tray that has sufficient volume to hold all the contents in the event of breakage.

Nitrates of silver or mercury should not be mixed with organic solvents; for example, alcohols, because explosive fulminates may be formed.

Hydrofluoric Acid

Hydrofluoric acid is one of the most active chemicals known. Few materials will withstand its action. The anhydrous acid will attack glass, leather, natural rubber, and most organic materials; it will rapidly corrode some metals, especially those containing silicon, such as cast iron and has a great affinity for water.

Burns from HF may be placed in three classes: (1) those resulting from concentrations up to 20%, which manifest themselves several hours after exposure by a deep-seated reaction; (2) those from concentrations of about 20-50% in which the latent period is shorter; and (3) those from concentrations of about

60% in which the burn is felt shortly after or immediately upon exposure. Flushing with copious amounts of water is the immediate first-aid treatment for burns. After thorough washing with water, burns should be treated at a medical facility.

Porous materials, such as concrete, wood, pipe coverings, and plaster, to name a few, which have been in contact with HF or involved in spills, will absorb the acid with consequent hazard for an indefinite period of time unless neutralized with a good agent, such as sodium carbonate.

Wash tools that have been in contact with this acid carefully with water after each use.

Operate all valves on equipment containing anhydrous HF from behind a barricade.

Before any maintenance work is started, any HF lines or equipment requiring repair must be carefully drained, flushed, and cleaned of all acid residues.

Fluorine

Fluorine is one of the most reactive materials known. Under favorable conditions it will attack practically all chemical elements. Fluorine can be kept in metal containers only because a protective layer of metal fluiride forms on contact and thus prevents further attack.

The occupational exposure limit is 1.0 ppm (tolerable for steady exposure every 8 hour working day). Human tests indicate that individuals can smell fluorine at nine parts per billion and that some can detect fluorine at three and one-half parts per billion. The odor of fluorine is persistent, but the sense of smell is not diminished by exposure. Where the odor persists, personnel should leave areas immediately and report the condition to the appropriate authorities. Respiratory effects of fluorine can appear as late as four hours after inhalation of quantities that can be described as "irritating"—usually more than 10 ppm. Some individuals report respiratory irritation at less than 10 ppm. Fifty parts per million is intolerable for anyone.

Fluorine gas attacks the skin directly, causing severe burns at high concentrations and irritation at lower levels. The eyes are particularly susceptible; the cornea can be damaged by brief exposure at quite low concentrations. Internally the first effect is irritation of the nose and throat. Heavy exposure may cause pulmonary edema (accumulation of fluid in the lungs). Chronic low-level exposure causes a variety of effects and must be safeguarded by regular medical examinations.

Heavy exposure of skin to 30 ppm of fluorine gas will cause a skin reaction after 15-30 minutes. The skin is damaged as the result of fluorine reacting with the moisture on the skin causing effects similar to a sunburn, which do not appear until several hours after exposure. Skin reactions will vary depending on the skin exposed, on the humidity, and on the type of work being performed.

Treat a fluorine burn on the skin as a combination chemical and thermal burn. If the exposure is slight, several hours may elapse before a person is conscious of pain. The exposed area becomes reddened, then swollen and pale, accompanied by a severe throbbing pain. Adequate treatment will usually stop pathological changes at this stage; without treatment, necrosis and ulceration will result. First-aid treatment should include immediate irrigation of affected external areas, including eyes, with a copious and prolonged supply of water.

The following safety measures are required for any operation involving the use of fluorine.

Fluorine safety measures

General. Any operation involving fluorine must be performed by at least two persons. An eye-wash fountain and safety shower must be located in the area.

Indoctrination. Each employee working with fluorine should understand the:

1. nature and properties of fluorine, especially its toxicity and reactivity
2. proper containers and methods of construction
3. suitable protective equipment and conditions of use
4. self-aid and first aid for fluorine inhalation and burns

Safety procedures. All personnel must wear safety glasses while in the laboratory.

All working areas in which fluorine is handled shall be outdoors or shall be ventilated by exhaust systems, hoods, blowers, etc., so that any accidental release of fluorine into the working area will be quickly removed.

Safety measures for specific conditions

Condition. Fluorine pressure in apparatus and connecting lines.

Safety measures. Use only approved components (tubing, fittings, valves, gauges, etc.) appropriate for the maximum fluorine temperature and pressure.

Condition. Fluorine at full supply tank pressure in apparatus and connecting lines.

Safety measures. Apparatus and connecting lines must be located behind a barricade to prevent direct exposure to a fluorine leak. All valves to be manipulated remotely, for example, by extending handles.

Condition. Working on fluorine apparatus and connecting lines.

Safety measures. Apparatus and connecting lines shall be evacuated or thoroughly purged with inert gas prior to opening. Personnel shall wear neoprene gloves when handling, dismantling, cleaning, etc. fluorine apparatus and connecting lines.

Condition. Accidental release of fluorine into working area.

Safety measures. All personnel shall immediately evacuate working area. Re-entry shall be made only after fluorine odor has been dispelled by area ventilation systems.

Medical surveillance

Do not assign any employee suffering from asthma or other pulmonary complaint, or from cardiovascular disease, or whose chest x-ray indicates an abnormal condition to operations involving fluorine.

An employee assigned to work with fluorine must submit to periodic chest x-ray examinations as prescribed by the appropriate medical authority.

Notify a physician immediately when any employee is exposed to fluorine—especially if fluorine has been inhaled.

Laser dyes

Laser dyes require special safety consideration. The labeling system for laser dyes gives little or no warning of associated hazards. Yet studies show that many dyes are extremely toxic by themselves and that dimethyl sulfoxide, a common solvent for such dyes, can intensify the dangers by transporting the dyes through the skin.

Complexity of the toxicity problem is compounded by the insolubility in water of the cyanines and carbocyanines; the cyanines generally are dissolved in dimethyl sulfoxide (DMSO), a solvent that is dangerous in itself as well as because of its ability to facilitate the transfer of molecules through biological membranes. Keep information on the toxicity and carcinogenic aspect of dyes updated.

Fire Prevention

Good housekeeping and proper precautionary techniques are the best protection against the incident of laboratory fires.

Remove all refuse from floors and tables; clean up combustible chemicals and solvents immediately after any spillage; and discard all waste properly.

Each employee should be familiar with the fire exteinguishers available in all buildings and laboratories.

Open flames from gas burners must be attended continuously, and rubber tubing should be inspected periodically for condition and fit. Do not leave pressure on a gas burner or torch, and clamp all connections.

Do not pour volatile liquids down a sink or drain.

To encourage fire prevention, install covered *metal waste cans* or treadle-operated containers. "Flame Tamer" covers on wastebaskets are recommended.

When filling a bottle with chemicals, allow enough space to take care of the hydrostatic expansion of the liquid, and provide permanent labels.

The quantity of ordinary flammable solvents, where use of the material is infrequent, should not exceed one-half liter (about one pint) in a metal container if purity is not paramount. In case of frequent use of a solvent such as ether, the solvents should be stored in a refrigerator in tightly closed metal cans or containers. Refrigerators used for the storage of flammable solvents should be rendered explosionproof by the removal of all electrical switches and light sockets.

When a mechanical stirring device must be used for a flammable solution, the stirrer must be activated by a water or air motor or an explosionproof electric motor.

Obvious fire hazards include: exceeding the flash point of liquids, allowing the accumulation of materials that might ignite spontaneously, and failure to remove flammable wastes.

Report fires immediately to the Fire Department by activating a fire-alarm box or by dialing the area emergency number (i.e., 911).

If a fire extinguisher has been used, notify the Fire Department immediately so that they may replace or refill the extinguisher.

Become familiar with the locations and the use of fire-alarm boxes, fire extinguishers, hose lines, safety showers, exit doors, and so on in your area.

Notify the Safety Office of all fires, no matter how small.

GENERAL SAFETY INFORMATION

Do not try to operate any equipment unless you are familiar with its operation and have been authorized to do so.

Do not work with power machinery unless another person is in the vicinity to render aid, if necessary.

Wear safety glasses, goggles, or face shields where machine tools are being operated.

Compressed air lines for general shop use should not carry more than 30 psi pressure, unless fitted with a safety nozzle approved for higher pressures. Do not use compressed air to blow dirt from machinery, clothing, or hair.

Do not attempt to remove foreign objects from the eye or body if injury is suspected. Contact a physician immediately.

Do not wear neckties, gloves, loose clothing, or long sleeves around moving machinery.

Remove finger rings when working with machinery or electricity.

Do not remove belt guards or other safety guards or make them ineffective.

Shut down all machines for cleaning, oiling, or repairing.

Use a brush, special tool, or hook to remove chips, shavings, or other material from work areas. Never use hands.

Keep fingers away from machines. Use special tools or devices, such as push sticks, hooks, pliers, etc. when necessary.

Keep the floor around the machine clean, dry, and free of tripping hazards.

Be careful when using a hard hammer to strike tool or machine parts; use a soft-face hammer, or a wood block, if possible.

Do not use a screwdriver as a chisel or lever.

Hand Tools

Choose good quality hand tools specifically for the job. Maintain them in good condition, use them correctly, and store them safely. Qualified personnel only will dress, temper, or make simple mechanical repairs on a hand tool. Send defective tools to salvage unless it is simple and safe to repair them.

When using tools on elevated surfaces, such as mezanines or platforms, take special precautions to prevent dropping them from these levels; for example, use kickplates.

Do not apply a screwdriver or other sharp tool to an object held in the hand. If possible, secure the object in a vise.

Use files with handles whenever possible.

Inspect chisels, punches, and similar tools frequently for evidence of mushroomed heads or burrs, if necessary, and, if necessary, machine dress.

Inspect wood-handled tools frequently for splitting, cracking, and splintering. Do not tape defective handles, send them in for repair.

Portable Power Tools

Portable power tools are sources of safety hazards, often exerting more force than hand tools of the same type and may present the hazard of electric shock.

Personnel should learn how to use the equipment safely and to inspect it before use to detect unsafe conditions.

Power tools should be grounded. However, tools rated as "double-insulated" by the Underwriter's Laboratories may be used instead of the grounded type.

Eye protection is required when using power tools.

Remove any hold-down buttons on hand tools, for example, drills.

Turn power off before installing, adjusting, or removing guards or other accessories. Leave guards in place while using the tool.

Place soldering irons in holders when not in use. This will eliminate the danger of fire and also protect personnel from accidental burns.

Disconnect power cords to soldering irons and guns when the soldering operation is completed, or at the end of the shift, whichever comes first.

Machine Operations

Drill press operations

Safety glasses, goggles, or face shields are required.

When drilling, tapping, or reaming, ensure that the work piece is securely fastened in a heavy vise by blocks or by clamps so that it cannot spin or climb the drill. Do not rely on hand strength to secure the material from turning.

After tightening a drill bit in the check of a drill press, remove the release key before starting the machine.

Become familiar with proper speeds for various types of drilling operations. Forcing or feeding too fast may cause broken drills and result in serious injury.

Do not loosen the chuck or a tapered shank unless the power is turned off.

Before removing chucks from the spindle, lower the spindle close to the table so that the chucks will not fall.

Do not remove drillings from work with your fingers.

Grinding operations

Safety glasses, goggles, or face shields are required.

Provide grinding wheels with tool rests that are set no more than 1/8 inch from the stone. Do not use the side of the emery wheel for heavy grinding unless it is of a special type for that purpose.

Stand to one side when starting up a machine, and do not exert great pressure on the wheel until it has attained correct operating speed.

Report to your supervisor immediately any broken, cracked, or otherwise defective wheel.

Experienced personnel *only* will mount a new wheel after ensuring that the wheel is of the proper rpm rating for the machine.

Circular saws

Safety glasses, goggles, or face shields are required.

Do not stand directly in the line of work being fed through a saw. Stand to one side.

Do not use a rip saw for cross-cutting; do not use a cross-cut saw for ripping.

Ensure that the saw is in good condition before using it, that is, with sharp, unbroken, uncracked, and properly angled teeth.

Do not reach over the saw to obtain material from the other side.

Do not oil the saw or change the gauge while the machine is running.

Do not let material accumulate on the saw table. Keep the floor free from slipping and stumbling hazards.

When shutting off power, do not thrust a piece of wood against the saw to stop it. Be sure the saw has stopped before leaving it.

Use a pusher stick whenever the size or shape of the piece requires the hands to be near the blade of the saw.

Leave the appropriate guards in place at all times.

Incorporate the antikick feature on circular saws.

Band saw operations

Safety glasses, goggles, or face shields are required not only for sawing operations but also if the operator is arc-welding to join or repair blades as provided by some models.

Keep adjustable guards as close to the point of operation as the work permits.

If a band breaks during operation, shut off the machine and stand clear until the machine has stopped.

Do not stop a machine by pushing material against the band.

Do not use cracked saw blades. A "click" as the blade passes through the work denotes a cracked blade or a missing tooth. Replace the blade before using the machine again.

Lathe operations

Personnel must not operate a lathe until they are familiar enough with its operation to perform the work safely. The book *How to Run a Lathe* (South Bend Lathe, Inc.) is recommended reading for all potential operators.

To protect against flying metal chips, use safety glasses, goggles, or face shields.

Avoid contact with faceplate projections, chucks, and lathe dogs, particularly the ones with projecting set screws.

When filing, hold the elbow high with the sleeves rolled up to prevent contact with the lathe dog. Files designed to take handles should also have one to prevent hand injuries.

After the chuck is adjusted, remove the chuck wrench immediately.

Do not use the hands to remove chips; use a brush.

Soldering and Welding

Silver soldering

Some silver solders contain as much as 18-20% cadmium. Due to the toxicity of cadmium, silver soldering should be pefformed in well-ventilated areas, under an approved exhaust hood, or while wearing an approved respirator. In addition, because nitric-oxide fumes from cleaning acids are poisonous, the Safety Office must approve all silver soldering operations.

Welding

Welders and their helpers must wear welding glasses or shields of proper shade during welding operations described in an approved standard work permit (SWP). Use of equipment for welding or torch cutting requires a fire watch.

Avoid handling oxygen containers with greasy hands or greasy gloves. Explosions have resulted from the reaction of oxygen with organics.

Valves of cylinders containing oxygen are designed to "back seat" and should be opened *fully* during use.

Mount H-size cylinders on trucks or strap them to a fixed object to prevent falling.

Inspect hoses, torch tips, and attachments before each use to ensure good working condition.

Use wrenches or tools provided or approved by the gas manufacturers to open the cylinder valves. Do not force open the valve wheel with a tool. Return cylinders with faulty valves to stock.

Shut cylinder valves off when the job is completed or when work is interrupted for longer than 30 minutes. Release the pressure from the regulator by opening the torch valves momentarily.

Do not weld any container that has held flammable materials such as gasoline, paint thinner, enamel, varnish, and so forth unless such containers have been carefully cleaned or filled with water.

During electric arc welding, set up a shield to prevent flash burns and flying sparks.

Ensure that all trash has been removed from the vicinity of welding or cutting operations and that combustible matter has been removed or shielded with flame-resistant guards or covers.

Keep suitable fire extinguishing equipment nearby and know how to operate it.

Materials Handling, Including Forklift and Hoist Operation

Special crafts conduct most of the rigging operations for delivery and installation of heavy material or equipment; however, many material handling requirements must be completed by other personnel not specifically trained for the hazardous manipulations of heavy objects. The majority of reported injuries and accidents occur in materials handling, therefore, educational courses for inexperienced personnel are recommended to provide training before assignments to duties in this field. Certification of operators of forklifts, hoists, and cranes is required as noted below.

Ropes, chains, slings, and other support accessory items are susceptible to damage if not handled and stored (hung) properly. Damage such as broken strands, nicks, scratches, tears, wear, to name a few may be observed and can reduce the load limits marked on the tags.

Manual handling

Inspect materials to be handled carefully for sharp or jagged edges, burrs, splinters, etc., and use protective devices to avoid injury.

Most back injuries are incurred during picking up or setting down materials. When lifting an object from the floor or setting it down, keep the back straight and lift with the legs. Get help if the object is too heavy for one person.

Hand carts

Two-wheeled and four-wheeled carts are valuable aids in materials handling, but take certain precautions to ensure safety in operation.

Load carts properly; stack the load evenly with heavy objects on the bottom.

Keep a cart ahead of you when going down an incline and behind you when going up an incline. If possible, do not walk backward with a two-wheeled cart.

Push four-wheeled trucks; *pull* those with handles.

Do not use special trucks designed for handling compressed gas cyliners for any other purpose.

When not in use, keep hand carts in a location where they do not present a hazard. Park them with handles in a vertical position.

Forklift truck

Operators must be certified by an approved instructor.

Some general rules and reminders for the safe operation of forklift trucks are listed below.

1. Do not speed.
2. Come to a complete stop before moving through a blind intersection.
3. Always look before backing up.
4. Load the truck in a manner that does not obstruct your view.
5. Stay within the load limit clearly marked on the truck.

6. When carrying a load, drive forward up a ramp and drive backward down a ramp.

7. Allow no riders.

8. The forklift truck is steered by the rear wheels; use caution when turning as well as when moving forward or backward to avoid collision accidents with pedestrians, materials, or equipment.

9. Keep forks in the lowest possible position when the engine is turned off, when the lift is not in use, and in as low a position as practical when transporting a load.

10. Do not play games with the forklift.

11. Normally, a forklift truck may not be used as a personnel elevator. Exceptions may be permitted by authorized personnel in locations or for situations where use of the forklift is the safest and most reasonable method. Use an appropriate cage or platform securely attached to the forks, and do not move the truck when the cage is in the elevated position.

12. When leaving the truck unattended, shut off the power, neutralize the hoist and gear controls, and set the parking brake. When not in use, park gasoline-driven lifts outside the buildings at least 10 feet from combustibles. Do not refuel while the motor is running or inside a building. Spilled fuel must be removed and the fueltank cap replaced before starting the engine.

13. Report all accidents to the supervisor immediately.

14. Unauthorized personnel must not perform maintenance work.

Jacks

Use a jack of proper capacity for the load. Each jack will have its capacity rating clearly displayed.

Do not subject a jack to extremely rough treatment, which may weaken it and cause it to fail under load.

Jacks are designed to be used in the vertical position. If used at an angle, take extra precautions to prevent slipping and swaying of the load.

Do not depend on a jack as the sole support to hold a load in an elevated position. Place substantial blocking under the load once it has been raised to the desired height.

Electric hoists and cranes

Operators of hoists and cranes must be certified.

Safe load capacity must be displayed clearly.

Pick up the load only when it is directly under the hoist.

Do not permit personnel to stand or walk under the hoist.

Do not permit riders.

Avoid dragging accessory chains, cables, or ropes across the floor.

Keep all guards and limiting devices in place during operation.

Exercise care to avoid picking up a load greater than the rated capacity of the hoist or crane.

Inspect hoists and cranes periodically.

Ropes

Inspect new ropes before using. Examine the entire length of rope, including inner fibers, frequently, and attach load limit tags.

Do not kink or twist rope.

Do not expose rope to rough treatment such as dragging across sharp or rough surfaces or positioning on sharp edges during lifting.

Splice, rather than knot, to join two pieces of rope permanently.

Do not use wet rope or rope reinforced with metal strands near power lines or electrical equipment.

Use manila, hemp, or nylon ropes, not cotton ropes, for lifting.

Wire rope

Inspect wire rope (cables) frequently for wear, broken wires, kinking, nicking, and lubrication; attach load-limit tags.

Apply noncorrosive lubricant regularly to wire rope used for hoisting operations.

Experienced personnel only will make eye splices. Use double clamps to prevent slippage.

Chains

Tag each chain to indicate its maximum safe load. If slings are at angles, consider the increased load.

Inspect chains for defects.

Do not splice chains by inserting a bolt between two links. Use replacement links available for this purpose.

Do not strain a kinked chain. Take up slack slowly to ensure that each link seats properly.

Use chain attachments of the same material as the chain.

Where practical, use chain hooks with safety latches.

Make sure the load is properly set in the bowl of the hook.

Order chain slings preferably from the manufacturer.

Slings

Label all slings with their maximum safe load.

Wire rope is more satisfactory for slings than chain or fiber rope.

When lifting a load, keep the legs of the sling as vertical as possible.

If practical, use safety hooks on slings.

Inspect jacks, hoists, cranes, ropes, chains, and slings for obvious defects before using. In addition, periodic safety inspections should be held in order to look for obvious defects.

Schedule detailed inspections of hoisting equipment at time intervals on the basis of use. The supervisor should keep records of the inspections and should request additional inspections of all lifting devices, when necessary. The use of equipment should be suspended until the inspection is made and the item or installation deemed satisfactory for further operation.

High-Pressure Apparatus

The hazards in high-pressure work are caused by (1) flying fragments from failure of high-pressure vessels, fittings, or gauges; (2) failures of high-pressure lines and hoses, valves, and connectors; (3) toxic effects; (4) secondary gas explosion; or (5) fire resulting from the release of gases when such a failure occurs. All equipment must be adequately designed, tested, maintained, and operated; location must be approved so that effects of a failure will be minimal.

Each item of high-pressure equipment should be designed to take full account of the effects of the subjected pressure. Examine vessels for cracks inside and out, and test them for leaks or weak spots by a suitable hydraulic procedure; then test the vessel at 125 or 150% of the maximum allowable working pressure, pneumatically or hydrostatically, respectively. If leaks are observed, release the pressure, repair the leaks, and repeat the checking process until the system is satisfactory. The American National Safety Institute (ANSI) codes provide a basis for test and inspection procedures, and copies of these codes are available from the Safety Office. Do not tighten leaking valves and connections under pressure. Leaks may result from erosion in the connection, which may cause the leak to increase rapidly when tightened. Cracked or split gaskets or packing rings may cause leaks and tightening the joint may cause a serious blowup.

Use barricades to protect personnel from injury and to prevent damage to other facilities. Fit high-pressure equipment adequately, with provisions for the release of unexpectedly high pressures. In small systems, provide this release in tubing connections away from the main vessel; but in large systems keep the openings for relieving pressures as close to the largest volume of gas or fluid as possible. Two types of pressure-relief devices in common use are the spring-loaded safety valve and the rupture disk.

High-pressure valves have at times caused accidents and deserve consideration of both location and operation. Consider valves as part of high-pressure equipment, and install them behind barricades with the stems pointing up and away from operators, if possible.

When high-pressure tubing or flex-lines fail, they whip about, causing possible damage and serious injury to personnel. Anchor tubing and lines securely to prevent this.

Bourdon pressure indicators have failed. Corrosion, metal fatigue, and embrittlement have led to these gauge failures. Drill pressure-relief holes (if not already present in the gauge case) in the cases, and replace the glass faces with plastic.

Compressed gas containers are considered high-pressure apparatus and safe for this purpose. Abuse in handling, use, and storage can cause serious accidents. Label all cylinders immediately.

Based on accident prevention experience within these industries, the Compressed Gas Manufacturer's Association has compiled the following rules, which should be observed in the handling of compressed gases.

1. Do not drop cylinders or permit them to strike each other violently.
2. Do not use a lifting magnet or a sling (rope or chain) when handling cylinders. Use a crane when a safe cradle or platform is provided to hold the cylinders.
3. Use the cylinders approved only for use in interstate commerce for the transportation of compressed gases.
4. Where caps are provided for valve protection, keep such caps in place except when the cylinders are in use.
5. Do not tamper with the safety devices in valves or cylinders.
6. Make sure that the threads on regulators, or other auxiliary equipment, are the same as those on the cylinder valve outlets. Do not force connections that do not fit.
7. Do not attempt to repair or alter gas cylinders or valves.
8. Do not use grease. oil, or other organic substances as a lubricant on valves, fittings, or gauges of cylinders containing oxygen.
9. Chain or strap cylinders securely in place so they will not fall over. Use cylinder support rings for extra tanks.

Give special attention to *combustible* compressed gases.

1. Keep sparks and flames away from cylinders.
2. Keep connections to piping, regulators, and other appliances tight to prevent leakage. Where using a hose, keep it in good condition.
3. Do not use an open flame to detect combustible gas leaks. Use soapy water or commercial leak-checking solutions. During freezing weather, use linseed oil.
4. When cylinders are not in use, keep valves tightly closed.
5. Reduce combustible gas cylinder pressures preferably through a pressure regulator attached directly to the cylinder.
6. After removing the valve cap, open the valve slightly for an instant to clear the opening of dust and dirt.
7. If a valve is difficult to open, return the cylinder to stock with an appropriate tag or label.
8. After attaching the regulator and before opening the cylinder valve, ensure that the adjusting screw of the regulator is released.
9. Do not permit the gas to enter the regulator suddenly. Open the valve slowly.
10. Before the regulator is removed from the cylinder, close the cylinder valve and release all gas from the regulator.
11. Use manifolds for combustible gases only if the manifolds are designed by qualified engineers.
12. Do not use combustible-gas regulators, hoses, or other appliances for other gases.
13. Do not store reserve stocks of cylinders containing combustible gases with cylinders containing oxygen. Group them separately.

Cryogenics

The major cause of cryogenic mishaps is the result of the expansion of liquids to gas state at boiling point. Other hazards associated with the use of cryogenic liquids are frostbite, explosion, fire, material failure, and toxicity. The extent of these hazards depends on both the physical and chemical properties of the individual cryogen.

Special instructions are required for users of cryogenics and a thorough study of the potential hazards is recommended.

Protective Equipment

A variety of personnel protective equipment should be made available to all employees. This includes *safety glasses* and all types of goggles of the nonprescription type. Employees who require prescription safety glasses are to be supplied with glasses at no cost to the employee other than the prescription.

Wear heavy rubber *gloves* of the gauntlet type for handling bulky, rough, or sharp materials, such as boxes and crates, metals, glass, or nonelectric cables. Employees engaged in welding, cutting, brazing, or working with hot metals should wear leather or asbestos gloves or aprons, depending on the type of work being done and on the exposure. Gloves should fit snugly above the wrist and, if necessary, should be tied so as to prevent acid splashes, sparks, and so on from entering the glove opening. Under most circumstances, gloves should not be worn close to revolving machinery or equipment, nor should metal-studded or riveted gloves be worn when work involves electricity or electrical apparatus.

Various types of *shoes* are available to employees in approved areas.

Wear *ear plugs* and muffs for protection against noise. Ear plugs must be fitted per instructions from a physician.

Respiratory protection equipment may be required to prevent exposures of personnel to toxic materials or oxygen-deficient environments. An industrial hygienist or a health physicist must determine the need for and the type of respiratory protection for each specific job. Medical certification is required before using any type of respiratory protective equipment, stating that the individual is physically able to perform the assigned duties while wearing such equipment. Individuals requiring the use of respiratory protective equipment must be trained in the use, care, and limitations of such equipment and must be individually fitted to assure that the best possible protection is afforded. Respirator users must be retrained, refitted, and medically approved every two years.

The *supervisor* is responsible for:

1. Training and fitting personnel who are required to use air-purifying devices, supplied air devices, or a self-contained breathing apparatus (SCBA)
2. Selecting and issuing appropriate respiratory protective equipment
3. Conducting quality assurance tests on the equipment to ensure that users are performing routine (usually monthly) checks of the equipment
4. Rechecking individual respirator fittings
5. Reviewing standard operating procedures (SOPs) that require employees to wear respiratory protection
6. Maintaining appropriate records of fitting/training/quality assurance data

A *physician* should be responsible for the medical review and approval of employees who use respiratory protective equipment and for informing the employee of any changes in physical capacity that may negate previous medical approval.

Employees must use proper care when using respiratory protective equipment and must notify their supervisor of any change in health that may affect performance while wearing respiratory protection. Use of SCBA or supplied air requires an SOP, which the employee must follow.

Limited Egress/Confined Spaces

Limited egress/confined spaces (LE/CS) constitute a major hazard to the safety of personnel who must work within such spaces. Generally, LE/CS are enclosures which have limited openings for entering and/or exiting, may lack adequate ventilation, or may contain life-endangering atmospheres due to the presence of toxic, flammable, or corrosive contaminants, or lack of sufficient oxygen. LE/CS may include, but is not limited to, storage tanks, process vessels, stacks, pits, degreasers, reactor vessels, boilers, ventilation and exhaust ducts, manholes, and any open-topped space more than four feet in depth not subject to adequate ventilation. The configuration of the enclosure and the operations conducted within the enclosure determine if an LE/CS exists.

Supervisors of the operating units must identify all enclosures in work areas under their control that may be defined as LE/CS, and SOPs developed for each location.

Standard Operating Procedures (SOPs) and
Special Work Permits (SWPs)

A standard operating procedure (SOP) is an approved document that (1) describes the controls and procedures for conducting an activity or operation that involves potentially hazardous conditions; (2) identifies the responsible personnel and organizations involved; and (3) provides details of support services, equipment to be used, and actions to be taken in the event of an emergency.

A service work permit (SWP) is an approved document that lists the limiting conditions and precautions to be observed in performing a limited-term or one-time operation conducted under potentially hazardous conditions.

In determining potentially hazardous activities, line-management personnel must determine whether or not an activity is potentially hazardous and requires an SOP and an SWP.

Safety Training

Employers should provide health and safety training to enable employees to perform their assigned tasks in a safe manner. The function of such training is for employees to recognize the risks associated with their job activities and to make an informed judgment before assuming such risks. Instruction in health and safety procedures, operating procedures, and the recognition of potential hazards are important elements of this program. Required health and safety training is separate and distinct from training provided primarily for career advancement.

To implement this policy, supervisors must instruct their employees in specific job requirements and all applicable health and safety procedures. The responsible supervisor must complete a health and safety job evaluation for each employee. Specifically, required health and safety training courses should be commensurate with the employee's duties. Records will be maintained to reflect the current status of training received by each employee, and testing of individuals will be conducted as required to assess the effectiveness of the training program and the employee's understanding of the job-related health and safety requirements. Successful completion of the required health and safety training program is essential for working in certain areas and in specific job categories.

Medical Surveillance

Employers will provide a comprehensive occupational medical program to assist in promoting health maintenance of employees. The program shall consist of periodic medical evaluation of employees, dispensary services for occupational injuries and diseases, consultation on potential health-related problems, employee counseling, medical evaluation of work capacity of individuals, training and education of employees on occupational or personal health subjects, maintenance of medical records, and other occupational medical activities designed to improve the health of the worker.

14

Special Safety Rules, Policies, Safety Committee Organization, and Emergency Procedures

The following rules and policies should be established. *They are inflexible.* (Extracted from Los Alamos National Laboratory manual.)

HAZARDOUS WORK RULE

Any work considered to be hazardous must be attended by at least two persons and must be approved by a supervisor. This applies to all work with electrical equipment (particularly lasers), radioactive materials, and hazardous chemicals; for example, beryllium.

In general, supervisor approval is given by adoption of standard operating procedures (SOPs).

If radioactive materials are to be handled, notify a superior in advance so that a monitor can be present. The handling of toxic materials must be approved by a superior.

Casual employees are *not* permitted to work in laboratories or with experimental apparatus during nonworking hours unless such work is approved by a superior *and* a staff member is present to supervise. Office work is exempt from this policy.

FIRE PREVENTION MEASURES

Keep only minimum amounts of any volatile, flammable solvents in a laboratory, preferably not exceeding one-half liter of each type needed; store them in safety cans provided for this purpose. Notify the Safety Office before using volatile, flammable solvents in any laboratory equipment; and do not use them in any drybox or hood not equipped with adequate ventilation or in an inert atmosphere.

To keep laboratories as fire resistant as possible, obtain Safety Office approval for all wooden construction objects and cover them with flame-retardant coatings.

LIQUID HYDROGEN

The use of liquid hydrogen in any area will be permitted only after the Liquid Hydrogen Safety Committee has approved the location and quantities to be used and has reviewed the practices involved.

DISPOSAL OF HAZARDOUS CHEMICALS

Notify a supervisor if surplus chemicals must be discarded in quantities that cannot be handled in the laboratory or which do not lend themselves to such disposal; for example, cyanides, alkali metals, flammable liquids, and hydrides. Disposal forms are to be made available.

Use cardboard boxes, closed bottles, or jars for proper packaging. Exercise care in packaging chemicals for disposal to ensure that compatible chemicals are separated in case of accidental breakage of glass containers. The package must then be monitored for radioactive contamination, if appropriate, and labeled for disposal.

Very small quantities of most acids do not require special disposal; just flush them down an acid drain with appropriate amounts of water. When disposing of chemicals in large quantities, or in large and heavy containers, such as drums or carboys, the group involved should be prepared to offer assistance for disposal.

Use special care in handling cyanide waste. Deliver it to the appropriate area for special treatment and disposal. The toxic limit of hydrocyanic acid (HCN) is about 1 ppm, and any discharge to the acid sewer should not result in a concentration greater than this value. Whenever these wastes are discharged to the sewer, they should first be made highly alkaline to prevent subsequent formation of HCN. A discharge limit of 10 g of HCN per hour or 150 g per 24-hour period is suggested. It is realized that there will be occasions in which the above requirement cannot be met and larger quantities must be discharged to the sewer. In these cases, it is requested that authorities be notified so that proper measures can be taken at the treatment plant.

EXPLOSIVES

The Los Alamos National Laboratory and the Department of Energy require that an approved standard operating procedure (SOP) be provided for each operation involving explosives. Strict adherence to these procedures is fundamental to explosives hazard control. Do not attempt any operation or experiment involving explosives without written approval from the Safety Office.

NEW PROCESSES

Before starting the actual work, discuss all new processes, projects, and experiments not previously carried out with the appropriate authorities.

All engineering drawings for such new work and new installations require Safety Office approval. Engineering plans for modifying existing ventilation or new plans for existing ventilation also require approval.

DIVISION SAFETY ORGANIZATION
(SUGGESTED FOR LARGE ORGANIZATIONS)

Each division shall prepare, submit for approval, and thereafter carry out an internal program of safety organization, the object of which shall be to assist the division leaders and group leaders in maintaining a high degree of safe practices and safety consciousness among their personnel. While the detailed procedures to be adopted will obviously differ from group to group, they could include the organization of a group (or combination of groups) Safety Committee, a specifically identified chairman, meetings at regular and predicted times (e.g., monthly), and possibly formal presentations of some aspect of safety. Formal minutes (and/or recommendations, if appropriate) should be prepared and a copy sent to the division leader and to the Safety Office. A representative of the Safety Office should be invited to these meetings and should be expected to attend, if possible.

Maintenance or repair orders whose primary objective is to be related to the improvement of safety of operations will be so identified by conspicuous lettering to this effect on their face. Such orders will receive priority attention and, if approved, will be expedited for completion as rapidly as feasible.

The Safety Office shall make periodic unannounced inspections of all activities with regard to the safety or their operations and safety practices. Such inspections shall be conducted at least annually and at least quarterly in all areas where laser, chemical, physical, mechanical, or electrical operations are regularly carried out. The reports of these inspections, whether or not containing recommendations for change, will be given formally to the division leader with a copy to the group leader. When recommendations for changes are made, follow up of the execution will be made by the division leader within 30 days and they will be checked by the Safety Office at their next inspection.

The corporate manager or the director's office is to be notified of any disagreement or failure to comply with any of the above procedures.

In compliance with the above policies, the following organizational procedures should be established.

1. *A division safety committee* shall consist of group leaders or designated representatives of all groups. In addition, a representative of the Safety Office will serve as chairman of this committee. Other division staff members may be added to the membership. This committee will meet quarterly to determine the general safety policy of the division and to advise the division leader of the status and progress of division safety practices. Formal minutes of each meeting will be forwarded to the division leader and the Safety Office.

2. *Group safety committees* shall consist of members from the individual groups. Each group leader shall appoint the chairman and members of the committee, which shall hold quarterly group safety meetings to discuss division safety items and/or hear formal presentations on some aspect of safety. Before the meeting, when possible, the Group Safety Committee, accompanied by a Safety Office representative and a division safety representative, shall inspect all facilities on a quarterly basis. Formal minutes of the meeting and recommendations resulting from the inspection shall be prepared with copies for the division leader and the Safety Office, as well as for each appropriate member of the individual group.

EMERGENCY PROCEDURES

Anyone involved in any accident that results in personal injury should be immediately concerned with the *first aid* that can be rendered. Following the accident, but without delay, a complete report of the details should be made to proper authorities.

Personal Injuries

First aid

At the scene of any accident, (1) consider what immediate aid must be rendered, then (2) summon help. Telephone for any emergency (i.e., 911), and seek medical advice. While awaiting medical help, check for the following conditions.

Stoppage of breathing.   This may be caused by electric shock or inhalation of toxic gases, among other causes. Immediate action is vital. If the person has received an electrical shock, remove the victim from the source of the current; if the victim has been gassed, remove him or her to fresh air. (Keep your own safety in mind.) Start mouth-to-mouth artificial respiration. If you are alone, give five or six breaths and check for a heart-beat. If no heart-beat can be detected, start manual heart compression, and after 30 seconds switch back to artificial respiration. Continue alternating until the victim revives or is pronounced dead. (For a description of these techniques, see the section on "Heart-Lung Resuscitation Procedures.")

Hemorrhage.   In the vast majority of cases, bleeding can be most effectively controlled by pressure directly over the wound. Place a piece of the cleanest cloth available over the wound with other wadded-up cloth on top of this. Maintain firm pressure either by hand or with a bandage and without interruption until final treatment preparations are completed. In severe and unusual cases, pressure on an artery between the wound and the heart or the use of a tourniquet may be necessary, but the need for such procedures is rare. *A tourniquet is to be used only if an extremity is partially or completely severed.*

Acid or chemical burns.   Flood affected area with large quantities of water and continue for a considerable period of time. Wash eyes with water for at least 15 minutes before applying additional treatment. Do not attempt to neutralize the offending chemical with other chemicals.

Fire in clothing.   When nearby, use safety showers; if water is not available, wrap the person in a blanket or other material or roll him or her on the ground to smother the flames. If clothing adheres to burned skin, do not attempt to remove the clothing; cut it around the burned area.

Frostbite or cryogenic "burns."   Thaw out affected part in warm water of about 108°F (112°F maximum). Do not rub or apply pressure. Do not administer alcohol or tobacco.

Reporting

Accident reporting is necessary for the prevention of further accidents and to fulfill requirements of the Workmen's Compensation Law as a protection for both the employee and the employer. Any investigation or documentation required as a result of an injury will come only after necessary medical treatment.

The Safety Office requires submission of an Accident/Incident Report.

Serious accidents should be reported immediately (following state of emergency) by supervisory personnel to the Safety Office. "Serious accident" is defined as a fatality, an injury that will probably require hospitalization, involvement of several people, an occupational illness likely to result in lost time, an unusual radiation incident, a significant near miss, a safety system failure, or a fire judged to cause destruction in excess of $50,000.

Major fires must be reported immediately to the fire department (i.e., 911). Any fires brought under control by employees must also be reported so that the fire department can replace fire extinguishers, verify that the fire has been extinguished, make necessary reports, and (most important) be alerted so an apparent minor incident cannot develop into a serious fire.

Radiological incidents, which are incidents involving chemicals or toxic materials, are evaluated by specialists. Contact them.

Heart-Lung Resuscitation Procedures

If a person has stopped breathing or his heart has stopped beating, start heart-lung resuscitation at once. *If the person is not breathing*:

1. *Clear the throat.* Wipe out any foreign matter in his mouth with your fingers or with a cloth wrapped around your fingers.
2. *Place victim on his back.* Place the victim on a firm surface such as the floor or the ground, not on a bed or a sofa.
3. *Tilt the head straight back.* Extend the neck up as far as possible (this will automatically keep the tongue out of the way).
4. Open your mouth wide and place it tightly over the victim's mouth. At the same time pinch the victim's nostrils shut. An alternative is to close the victim's mouth and place your mouth over his nose. This latter method is preferable with babies and small children. *Blow* into the victim's mouth or nose with a smooth, steady action until the victim's chest is seen to rise.
5. *Remove mouth.* Listen for the return of air that indicates air exchange.

6. *Repeat.* Continue with relatively shallow breaths, appropriate for size, at the rate of one breath each five seconds.

If you are not getting air exchange, quickly recheck position of head, turn the victim on his side, and give several sharp blows between the shoulder blades to jar foreign matter free. Sweep fingers through the victim's mouth to remove foreign matter.

After four or five breaths, stop and check the pulse to determine if the heart is beating. If the heart is beating, return to the mouth-to-mouth resuscitation and continue until breathing starts or until a physician tells you to stop.

If the heart has stopped, begin heart massage:

1. *Place the heel of one hand on the lower third of the breastbone*, the other hand on top of the first.
2. *Thrust downward* from your shoulders with enough force to depress the breastbone 1½-2 inches.
3. *Relax* at the end of each stroke to permit natural expansion of the chest.
4. *Repeat* at the rate of about one per second.

If you are alone with the victim, you must alternate mouth-to-mouth breathing with heart massage at the ratio of about 2 to 15 (2 breaths, then 15 heart compressions).

If you have help, the ratio is 1 to 5. After five heart compressions, pause slightly to allow your partner to breath once into the lungs of the victim.

Call for help. continue one or both of the above while the victim is being transported to the hospital, or until he revives, or until told to stop by a physician.

15

Elements of a Successful
Laser Safety Program

The first concern in control of laser hazards is to become thoroughly familiar with all of the characteristics of the laser involved. The essential traits of the system are: (1) wavelength; (2) maximum available output power density, if continuous wave (CW), or maximum available output energy density and pulse width, if pulsed; (3) the classification per ANSI Z136.1; (4) the beam's skin-damage threshold value; and (5) the beam's eye-damage threshold value. These data will establish the required engineering controls and permit the selection of protective eyewear suitable for complete absorption of the beam or for controlled beam attenuation in case of beam detection—the most important single item in laser hazard control (see Fig. 15.1).

After learning the beam characteristics, the next step in becoming knowledgeable about the hazards of a laser is to assign safety responsibility to a laser safety officer to ensure adherence to requirements of the ANSI Z136.1 Standard, such as signs, lights, interlocks, enclosures, beam stops, and so on. Then indoctrinate all personnel who are to be assigned to the laser environment as to not only the potential hazards of the beam but also to any nonradiation hazards associated with the laser's operation. Two 16-mm movies are recommended for consideration of presentation to laser personnel: (1) "Lasers and Your Eyes," available from Los Alamos National Laboratory, Film Library, Box 1663, Los Alamos, New Mexico 87545, and (2) "High Voltage Angel,"

Figure 15.1 Photograph of laser personnel at the "front end" of a high power Nd:YAG laser system wearing two different styles and different lenses for eye protection. The eyewear lens material on the left is Schott's glass BG-18 and that on the right is Schott's glass KG-3 (with bifocals). (Courtesy of Los Alamos National laboratory, Los Alamos, New Mexico.)

available through Film Library, Lawrence Berkeley Laboratory, Berkeley, California 94720.

Personnel should be offered eye examinations per ANSI Z136.1 (Fig. 15.2) before being assigned to duties using the laser. The Laser Safety Officer should describe the choices of eyewear available to each individual since corrections may be required. Comfort and fit should be emphasized in recommending frames.

A written standard operating procedure (SOP) should be presented to each individual assigned to tasks in the laser environment with a sign-off sheet to show that the SOP has been studied and understood. The document need not be long and detailed, but salient safety procedures should be emphasized and responsibilities should be stipulated in clear and concise language.

Figure 15.2 Photograph depicting typical examination of eye of laser user before assignment to laser environment. (Courtesy of Los Alamos National Laboratory, Los Alamos, New Mexico.)

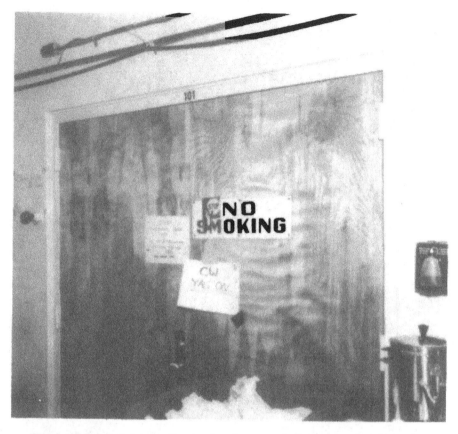

Figure 15.3 Photograph of entry to a laser research laboratory immediately before a laser safety program was instituted in 1972. (Courtesy of Los Alamos National Laboratory, Los Alamos, New Mexico.)

The LSO's duties should include a facility inspection on a regular basis, such as quarterly, followed by a written report that includes nonconformances to the ANSI Standard and to employer policy requirements. (Does entrance to your laser facility resemble the doorway shown in Fig. 15.3? Or is the entrance more like that shown in Fig. 15.4?)

By following these suggested procedures and by keeping safety precautions as an integral part of laser operations, accidents with laser systems can be prevented in this special field of high technology.

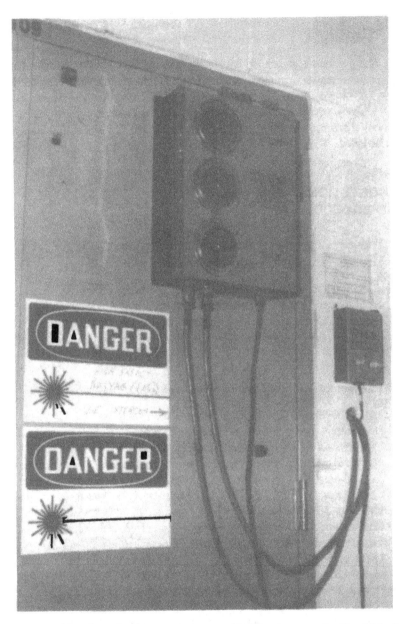

Figure 15.4 Photograph of entry to same facility as shown in Fig. 15.3 after a practical laser safety program was instituted. (Courtesy of Los Alamos National Laboratory, Los Alamos, New Mexico.)

233

This practical approach to laser safety resulted in no known permanent biological damage to any employees from laser radiation during the period 1972-1982 when the author was the Laser Safety Officer for the Laser Research and Technology Division of the Los Alamos National Laboratory, and dozens of lasers of a variety of characteristics were in use by as many as 400 laser personnel.

SUGGESTIONS FROM ANSI Z136.1, APPENDIX D

Guide for Organization and Implementation of Laser Safety and Training Programs

The extent to which the various parts of the following guide are applicable to a specific organization depends on the magnitude of the potential laser hazards within that organization. However, it is essential that each laser safety program include sufficient education of personnel in laser safety.

D1. Responsibility and Authority of Laser Safety Officer (LSO)

D1.1 General. The Laser Safety Officer (LSO) may be designated from among such personnel as the radiation protection officer, industrial hygienist, safety engineer, laser specialist, laser operator, etc. The LSO may be a user if there are very few laser operations or if the class of lasers and potential hazards are low. The LSO may be a part-time position when the workload for an LSO does not require a full-time effort. The LSO will effect the knowledgeable evaluation and control of laser hazards, and have the authority to monitor and enforce the control of laser hazards. In addition to the responsibilities described in section 1.3.2, the LSO's authority may include, but is not necessarily limited to, the responsibilities of operation specified in D1.2 through D1.8. The LSO will be trained in accordance with D6.

D1.2 Consultative Services. The LSO will provide consultative services on laser hazard evaluation and controls and on personnel training programs. Such evaluations will include electrical and other non-radiation hazards of lasers.

D1.3 Regulations. If a Safety Committee does not exist, the LSO may establish and maintain adequate regulations for the control of laser hazards.

D1.4 Authority. The LSO will have the authority to suspend, restrict, or terminate the operation of a laser or laser system if he or she deems that laser hazard controls are inadequate.

D1.5 Records. The LSO will ensure that the necessary records required by applicable government regulations are maintained. The LSO will also submit to the appropriate medical officer the individuals' names that are obtained in accordance with D4.4, D4.5, and D5.3. The LSO will ensure that the appropriate records are maintained indicating that applicable medical examinations have been scheduled and performed and that appropriate training has been provided.

D1.6 Surveys and Inspections. The LSO will survey by inspection, as considered necessary, all areas where laser equipment is used. The LSO will also accompany regulatory agency laser equipment inspectors, such as those representing OSHA, FDA/CDRH, state agencies, etc., and document any discrepancies noted. The LSO will ensure that corrective action is taken where required.

D1.7 Accidents. On notification of a known or suspected accident resulting from operation of a laser, the LSO will investigate the accident and initiate appropriate action. This may include the preparation of reports to applicable agencies.

D1.8 Approval of Laser System Operation. Approval of a laser or laser system for operation will be given. These include special operating procedures for maintenance and service operations within enclosed systems, and operation procedures for Class 3 and 4 systems. The procedures will include adequate consideration to assure safety from electrical hazards.

D2. Deputy Laser Safety Officers

If necessary, a Deputy Laser Safety Officer will be appointed by management. The Deputy Laser Safety Officer will perform the functions of the LSO when the latter is not available.

D3. Safety Committee

D3.1 Necessity. If warranted by the magnitude of the potential hazards of laser operations within the organization, a Safety Committee may be formed.

D3.2 Membership of Laser Safety Committee. The membership may include individuals with expertise in laser technology or in the assessment of laser hazards. Management may be included in the membership. Examples of members include, but are not limited to: technical manager, LSO (or other representatives from the safety department), physician and user representatives.

D3.3 Policies and Practices. The committee will establish and maintain adequate policies and regulations for the control of laser hazards and recommend the appropriate laser safety training programs and materials.

D3.4 Standards. The committee will maintain an awareness of all applicable new or revised laser safety standards. This may require liaison with regulatory agencies, either directly or through representatives.

D4. Responsibility of the Laser or Laser System Supervisor

D4.1 Prerequisite. The supervisor will be knowledgeable of education and training requirements for laser safety, the potential laser hazards and associated control measures for all lasers and laser systems, and all procedures pertaining to laser safety at locations under the supervisor's authority.

D4.2 Indoctrination. The supervisor will be responsible for the issuance of appropriate instructions and training materials on laser hazards and their control to all personnel who may work with lasers that are operated within the supervisor's jurisdiction.

D4.3 Laser Hazard Control. The supervisor will not permit the operation of a laser unless there is adequate control of laser hazards to employees, visitors, and the general public.

D4.4 Individuals Scheduled to Work with Lasers. The supervisor will submit the names of individuals scheduled to work with lasers to the LSO and, in addition, will submit information as requested by the LSO for medical surveillance scheduling and training completion.

D4.5   Reporting of Known or Suspected Accidents.   When the supervisor knows of or suspects an accident resulting from a laser operated under his or her authority, the supervisor will immediately notify the LSO or other designated authority.

D4.6   Medical Attention.   If necessary, the supervisor will assist in obtaining appropriate medical attention for any employee involved in a laser accident.

D4.7   Approval of Laser System Operation.   The supervisor will not permit operation of a new or modified laser under his or her authority without the approval of the LSO.

D4.8   Approval of Planned Installation.   The supervisor will make sure that plans for laser installations or modifications of installations are submitted to the LSO for approval.

D4.9   Operating Procedures.   For Class 3a, Class 3b or Class 4 lasers and laser systems the supervisor will be familiar with the operating procedures and ensure that they are provided to users of such lasers.

D5.   Responsibility of Employees Working with or near Lasers

D5.1   Authorizations.   An employee will not energize or work with or near a laser unless authorized to do so by the supervisor for that laser.

D5.2   Compliance.   An employee will comply with the safety rules and regulations prescribed by the supervisor and the LSO. The employee will be familiar with all operating procedures.

D5.3   Accident Reporting.   When an employee operating a laser knows or suspects that an accident has occurred involving that laser, or a laser operated by any other employee, and that such accident has caused an injury or could potentially have caused an injury, he or she will immediately inform the supervisor. If the supervisor is not available, the employee will notify the LSO.

D6. Training

D6.1 General. Training will be provided to each employee routinely working with or around lasers above Class 2. The level of training will be commensurate with the degree of potential laser hazards.

D6.2 Laser Safety Training Program Topics. Topics for inclusion in a laser safety training program may include, but are not necessarily limited to, the following:

1. For personnel routinely working on or around lasers:
   a. fundamentals of laser operation (physical principles, construction, etc.);
   b. bioeffects of laser radiation on the eye and skin;
   c. relations of specular and diffuse reflections;
   d. nonradiation hazards of lasers (electrical, chemical, reaction by-products, etc.);
   e. laser and laser system classifications;
   f. control measures;
   g. overall management and employee responsibilities;
   h. medical surveillance practices (if applicable);
   i. CPR for personnel servicing or working on lasers with exposed high voltages and/or the capability of producing potentially lethal electrical currents;

2. For the LSO or other individuals responsible for the laser safety program, evaluation of hazards, and implementation of control measures, or any others directed by management to obtain a thorough knowledge of laser safety:
   a. the topics in D6.2 (1);
   b. laser terminology;
   c. types of lasers, wavelengths, pulse shapes, modes, power/energy;
   d. basic radiometric units and measurement devices;
   e. MPE levels for eye and skin under all conditions;
   f. laser hazard evaluations, range equations, and other calculations.

Index

Milton Keynes UK
Ingram Content Group UK Ltd.
UKHW020027071024
449327UK00032B/2957